AF361901

Work Psychology and Occupational Health

Work Psychology and Occupational Health

Guest Editor

Masahito Fushimi

Basel • Beijing • Wuhan • Barcelona • Belgrade • Novi Sad • Cluj • Manchester

Guest Editor
Masahito Fushimi
Akita University Health Center
Akita University
Akita City
Japan

Editorial Office
MDPI AG
Grosspeteranlage 5
4052 Basel, Switzerland

This is a reprint of the Special Issue, published open access by the journal *International Journal of Environmental Research and Public Health* (ISSN 1660-4601), freely accessible at: www.mdpi.com/journal/ijerph/special_issues/_work_psychology_and_occupational_health.

For citation purposes, cite each article independently as indicated on the article page online and using the guide below:

Lastname, A.A.; Lastname, B.B. Article Title. *Journal Name* **Year**, *Volume Number*, Page Range.

ISBN 978-3-7258-3180-7 (Hbk)
ISBN 978-3-7258-3179-1 (PDF)
https://doi.org/10.3390/books978-3-7258-3179-1

Contents

About the Editor

Masahito Fushimi

Dr. Masahito Fushimi is the director as well as a professor at the Akita University Health Center, Japan. On March 31, 2016, he resigned from the position of director at the Akita Prefectural Mental Health and Welfare Center as well as from his work as a psychiatrist at the Akita Prefectural Center for Rehabilitation and Psychiatric Medicine. He was appointed the director and a professor at the Akita University Health Center on April 1, 2016. He works as a psychiatrist. He also serves as a counselor at the Akita Occupational Health Promotion Center. His main areas of interest in this field are mental health promotion, especially occupational stress management, and suicide prevention. His studies aim to provide a new understanding of the impact of stress on psychological and psychiatric aspects. Additionally, his research is directed at acquiring a better understanding of the environmental and psychological contributions to the etiology of stress-related disorders with a focus on mood and anxiety disorders and attempts to develop better means of prevention and treatment. His research has been published in several scientific journals. He has worked together with multidisciplinary clinicians and researchers to help prevent suicide through education. He has also participated in some research projects developed in Japan to investigate suicidal behavior. In addition, he has contributed chapters to several psychiatric books and written papers about suicidal behavior.

Preface

The COVID-19 pandemic has profoundly impacted workers' health worldwide, particularly their mental well-being. While many studies have highlighted the significant repercussions for workers' well-being, there remains insufficient evaluation and understanding of how the pandemic has affected employees and businesses, especially regarding psychological effects and coping strategies. This Special Issue explores the influence of the pandemic on the work environment, the relationship between it and workers' psychological states, and strategies for preserving workers' health—especially their mental health. The included studies stress the importance of acknowledging current conditions and adopting workplace and personal strategies to mitigate stress and build resilience throughout the pandemic. Key focus areas of this Special Issue are as follows: boosting telework and online operations; modifying the work environment through shorter hours, layoffs, and unemployment impacts; de-escalating stress levels paired with reduced coping strategies among employees; de-escalating stress caused by increased loneliness, isolation, and restricted communication opportunities (obtaining mental health resources); and implementing self-care practices for managing stress. This also involves practical approaches for addressing heightened stress and loneliness by providing robust support at the organizational and individual levels. This Special Issue, titled "Work Psychology and Occupational Health", includes contributions that explain impacts and applications of research and evaluations from various projects within this field context. This special issue provides valuable insights for conducting pertinent research and evaluations and guiding future studies. This collection features a variety of studies from a global group of authors, enhancing the perspectives included in this body of work.

Masahito Fushimi
Guest Editor

International Journal of
Environmental Research and Public Health

Editorial

Work Psychology and Occupational Health: An Editorial

Masahito Fushimi

Akita University Health Center, Akita University, 1-1 Tegatagakuen-machi, Akita 010-8502, Japan;
fushimi@gipc.akita-u.ac.jp or mfushimi8@gmail.com

Received: 3 January 2025
Accepted: 10 January 2025
Published: 13 January 2025

Citation: Fushimi, M. Work Psychology and Occupational Health: An Editorial. *Int. J. Environ. Res. Public Health* **2025**, *22*, 100. https://doi.org/10.3390/ijerph22010100

Globally, the COVID-19 pandemic has severely impacted workers' health, particularly their mental well-being [1–4]. While numerous studies have highlighted significant consequences for workers' well-being, there remains a lack of thorough evaluation and understanding of the pandemic's effects on employees and businesses, particularly concerning the psychological impacts and coping methods [5–10]. We are thrilled that this Special Issue includes contributions examining the impact of the pandemic on the work environment, the connection between the pandemic and workers' psychological effects, and strategies for maintaining workers' health—particularly mental health—during this time. These studies underline the necessity of recognizing the current circumstances and implementing workplace and individual strategies to alleviate stress and enhance resilience during the pandemic [Contributions 1–10].

Several papers in this Special Issue focused on essential topics such as healthcare workers' stress and psychological impact during the COVID-19 pandemic [Contributions 1, 2, 4]. Considering these factors, Bellehsen and his team explored how to modify the stress first aid model for healthcare workers on the frontlines during COVID-19 [Contribution 1]. They observed that introducing stress first aid could equip workplaces and individuals with essential skills to alleviate stress and foster resilience. These skills are beneficial during periods of healthcare-related stress. Discussing strategies for stress reduction and enhancing resilience is crucial for effectively designing and delivering healthcare services. Safiye et al. presented their cross-sectional study on mentalizing, resilience, and healthcare workers' mental health amid the COVID-19 pandemic [Contribution 2]. Their research importantly seeks to highlight key insights regarding the factors influencing depression, anxiety, and stress, emphasizing the critical need for developing and applying strategies that promote resilience and improve the mentalizing abilities of healthcare workers. Lee and colleagues present their exploratory study on nurses' emotions related to COVID-19 following their experiences with SARS [Contribution 4]. From discussions with nurses who experienced both SARS and COVID-19, it became clear that the media is a vital resource during disease outbreaks. Government departments need to exercise their expertise to utilize media channels effectively. For front-line nurses, it is crucial to grasp the public's reaction to the disease, receive on-the-job training and guidelines, and have adequate support administration.

The COVID-19 pandemic has had a psychological impact on workers other than healthcare workers, including in education and research [11–14]. Bearing this in mind, Padmanabhanunni and his team demonstrate in their study the factors leading to and the psychological impacts of teacher burnout during the COVID-19 pandemic [Contribution 3]. The authors highlighted that the dimensions of burnout are strong predictors of psychological well-being indicators, such as depression, hopelessness, anxiety, and life satisfaction. They stress that intervention strategies to lower burnout should equip teachers with sufficient job resources to help them cope with the demands and stressors of their roles.

Ramos-García and colleagues presented their exploratory analysis of university employees, examining the ergonomic factors affecting job satisfaction and occupational health during the SARS-CoV-2 pandemic through a structural equation model [Contribution 7]. They noted that organizational and physical factors positively impact job satisfaction. They determine that the findings enhance understanding of ergonomic factors and bolster educational institutions' sustainability efforts.

Moreover, the SARS-CoV-2 pandemic has highlighted the importance of considering the work environment as an ongoing influence, even after the crisis. This awareness has sparked increased interest in exploring new working arrangements, particularly remote work [Contribution 10], and research into the psychological impact on workers greatly affected by the pandemic, especially those involved in emotional labor [Contributions 5, 6]. Given this context, Cheung's study demonstrates the practical aspects of managing workplace well-being in post-pandemic remote work settings [Contribution 10]. He emphasizes that staff well-being is vital to managing organizational change. Senior management must implement sensible measures to improve occupational safety in their work-from-home (WFH) policies, following practical recommendations.

Additionally, he suggests that by adopting the proposed evidence-based practices for WFH initiatives, senior management can significantly enhance the overall workplace well-being of employees in the post-pandemic era. Cheng and colleagues delved into emotional labor with two contributions [Contributions 5, 6]. First, they examined how psychological empowerment mediates the relationship between transformational leadership and emotional labor [Contribution 5]. The authors emphasize that these findings enhance our comprehension of transformational leadership's impact on front-line employees' emotional labor. Additionally, they highlighted that the perceived external prestige of these employees affects their emotional labor via organizational identification and impression management motives, with the influence of each pathway varying based on the perceived organizational support [Contribution 6].

Several of our studies specifically examined the recommendations for mental health professionals and education policymakers to tackle the issues related to educational leadership. In particular, the case study by Wang and colleagues investigated the challenges faced by academic leaders, highlighting the essential factors required to ensure a safe workplace [Contribution 8], and the case study by Zhang and colleagues explored leadership [Contribution 9]. These topics are crucial for workplace mental health even after the pandemic. Wang and colleagues presented a case study examining occupational stress among doctoral supervisors in Chinese higher education institutions [Contribution 8]. They offered their recommendations on how mental health professionals and educational policymakers can tackle the concerns of doctoral supervisors by analyzing stress triggers. Zhang and team detailed the impact of safety management leadership on employees' safety performance based on a case study conducted in China's mining sector [Contribution 9]. They pinpointed key dimensions by exploring the correlations between leadership safety practices and safety outcomes.

In conclusion, the contributions to this Special Issue offer valuable insights for conducting research and evaluation within this context, informing future inquiries on the topic. This compilation features diverse studies from an international pool of authors, enhancing the perspectives presented in this body of work. I am especially pleased that early career researchers heeded our invitation to submit their work, highlighting our commitment to facilitating their publishing efforts. We appreciate the mentorship provided to them by seasoned colleagues and the support from the editorial team at IJERPH. I send a heartfelt thank you to every author who responded to our call for papers and our diligent independent peer reviewers who provided critical feedback. I also acknowledge the thousands of readers of this Special Issue. Lastly, I extend my gratitude to all those involved in the research highlighted here, whose contributions deepen our understanding of this vital field.

I look forward to potentially launching a second edition influenced by this collection, but for now, I hope you find joy in exploring this Special Issue.

Conflicts of Interest: The author declares no conflicts of interest.

List of Contributions

1. Bellehsen, M.H.; Cook, H.M.; Shaam, P.; Burns, D.; D'Amico, P.; Goldberg, A.; McManus, M.B.; Sapra, M.; Thomas, L.; Wacha-Montes, A.; et al. Adapting the Stress First Aid Model for Frontline Healthcare Workers during COVID-19. *Int. J. Environ. Res. Public Health* **2024**, *21*, 171. https://doi.org/10.3390/ijerph21020171.
2. Safiye, T.; Gutić, M.; Dubljanin, J.; Stojanović, T.M.; Dubljanin, D.; Kovačević, A.; Zlatanović, M.; Demirović, D.H.; Nenezić, N.; Milidrag, A. Mentalizing, Resilience, and Mental Health Status among Healthcare Workers during the COVID-19 Pandemic: A Cross-Sectional Study. *Int. J. Environ. Res. Public Health* **2023**, *20*, 5594. https://doi.org/10.3390/ijerph20085594.
3. Padmanabhanunni, A.; Pretorius, T.B. Teacher Burnout in the Time of COVID-19: Antecedents and Psychological Consequences. *Int. J. Environ. Res. Public Health* **2023**, *20*, 4204. https://doi.org/10.3390/ijerph20054204.
4. Lee, H.-L.; Chang, P.-J.; Lin, L.-C. An Exploratory Study of Nurses' Feelings about COVID-19 after Experiencing SARS. *Int. J. Environ. Res. Public Health* **2023**, *20*, 2256. https://doi.org/10.3390/ijerph20032256.
5. Cheng, P.; Liu, Z.; Zhou, L. Transformational Leadership and Emotional Labor: The Mediation Effects of Psychological Empowerment. *Int. J. Environ. Res. Public Health* **2023**, *20*, 1030. https://doi.org/10.3390/ijerph20021030.
6. Cheng, P.; Jiang, J.; Liu, Z. The Influence of Perceived External Prestige on Emotional Labor of Frontline Employees: The Mediating Roles of Organizational Identification and Impression Management Motive. *Int. J. Environ. Res. Public Health* **2022**, *19*, 10778. https://doi.org/10.3390/ijerph191710778.
7. Ramos-García, V.M.; López-Leyva, J.A.; Ramos-García, R.I.; García-Ochoa, J.J.; Ochoa-Vázquez, I.; Guerrero-Ortega, P.; Verdugo-Miranda, R.; Verdugo-Miranda, S. Ergonomic Factors That Impact Job Satisfaction and Occupational Health during the SARS-CoV-2 Pandemic Based on a Structural Equation Model: A Cross-Sectional Exploratory Analysis of University Workers. *Int. J. Environ. Res. Public Health* **2022**, *19*, 10714. https://doi.org/10.3390/ijerph191710714.
8. Wang, X. Occupational Stress in Chinese Higher Education Institutions: A Case Study of Doctoral Supervisors. *Int. J. Environ. Res. Public Health* **2022**, *19*, 9503. https://doi.org/10.3390/ijerph19159503.
9. Zhang, S.; Hua, X.; Huang, G.; Shi, X. How Does Leadership in Safety Management Affect Employees' Safety Performance? A Case Study from Mining Enterprises in China. *Int. J. Environ. Res. Public Health* **2022**, *19*, 6187. https://doi.org/10.3390/ijerph19106187.
10. Cheung, V.K.L. Practical Considerations of Workplace Wellbeing Management under Post-Pandemic Work-from-Home Conditions. *Int. J. Environ. Res. Public Health* **2024**, *21*, 924. https://doi.org/10.3390/ijerph21070924.

References

1. Lai, J.; Ma, S.; Wang, Y.; Cai, Z.; Hu, J.; Wei, N.; Wu, J.; Du, H.; Chen, T.; Li, R.; et al. Factors associated with mental health outcomes among health care workers exposed to coronavirus disease 2019. *JAMA Netw Open.* **2020**, *3*, e203976. [CrossRef] [PubMed]
2. Serafini, G.; Parmigiani, B.; Amerio, A.; Aguglia, A.; Sher, L.; Amore, M. The psychological impact of COVID-19 on the mental health in the general population. *QJM* **2020**, *113*, 531–537. [CrossRef] [PubMed]
3. Wang, C.; Horby, P.W.; Hayden, F.G.; Gao, G.F. A novel coronavirus outbreak of global health concern. *Lancet* **2020**, *395*, 470–473. [CrossRef] [PubMed]
4. Sadatnia, M.; Jalali, A.; Tapak, L.; Shamsaei, F. The Relationship between Mental Health and Loneliness in the Elderly during the COVID-19 Pandemic. *J. Ageing Longev.* **2023**, *3*, 433–441. [CrossRef]

5. Sher, L. The impact of the COVID-19 pandemic on suicide rates. *QJM Int. J. Med.* **2020**, *113*, 707–712. [CrossRef]

6. Sher, L. Psychiatric disorders and suicide in the COVID-19 era. *QJM Int. J. Med.* **2020**, *113*, 527–528. [CrossRef]

7. Fushimi, M. The importance of studying the increase in suicides and gender differences during the COVID-19 pandemic. *QJM Int. J. Med.* **2022**, *115*, 57–58. [CrossRef]

8. Bahrami Nejad Joneghani, R.; Bahrami Nejad Joneghani, R.; Dustmohammadloo, H.; Bouzari, P.; Ebrahimi, P.; Fekete-Farkas, M. Self-Compassion, Work Engagement and Job Performance among Intensive Care Nurses during COVID-19 Pandemic: The Mediation Role of Mental Health and the Moderating Role of Gender. *Healthcare* **2023**, *11*, 1884. [CrossRef] [PubMed]

9. Hincapié Pinzón, J.; da Silva, A.M.B.; Machado, W.L.; Moret-Tatay, C.; Ziebell de Oliveira, M. Transcultural Comparison of Mental Health and Work–Life Integration Blurring in the Brazilian and Spanish Populations during COVID-19. *J. Pers. Med.* **2023**, *13*, 955. [CrossRef] [PubMed]

10. Moodley, J.K.; Hove, R. Pastoral Care and Mental Health in Post-Pandemic South Africa: A Narrative Review Exploring New Ways to Serve Those in Our Care. *Religions* **2023**, *14*, 477. [CrossRef]

11. Fushimi, M. Student mental health consultations at a Japanese university and the current state of affairs on the increase in suicide victims in Japan during the COVID-19 pandemic. *Psychol. Med.* **2023**, *53*, 2174–2175. [CrossRef] [PubMed]

12. Weißenfels, M.; Klopp, E.; Perels, F. Changes in Teacher Burnout and Self-Efficacy During the COVID-19 Pandemic: Interrelations and e-Learning Variables Related to Change. *Front. Educ.* **2022**, *6*, 736992. [CrossRef]

13. Sánchez-Pujalte, L.; Mateu, D.N.; Etchezahar, E.; Gómez Yepes, T. Teachers' Burnout during COVID-19 Pandemic in Spain: Trait Emotional Intelligence and Socioemotional Competencies. *Sustainability* **2021**, *13*, 7259. [CrossRef]

14. Sokal, L.; Trudel, L.E.; Babb, J. Canadian teachers' attitudes toward change, efficacy, and burnout during the COVID-19 pandemic. *Int. J. Educ. Res.* **2020**, *1*, 100016. [CrossRef]

International Journal of
Environmental Research and Public Health

Essay

Practical Considerations of Workplace Wellbeing Management under Post-Pandemic Work-from-Home Conditions

Victor K. L. Cheung [1,2]

1   Department of Neuroscience, Psychology & Behaviour, University of Leicester, Leicester LE1 7RH, UK; cheungklv.iop@gmail.com
2   Multi-Disciplinary Simulation and Skills Centre, Queen Elizabeth Hospital, Hong Kong SAR, China

Abstract: As a natural experiment or "stress test" on the rapidly shifting work environment from office to home during and after the COVID-19 pandemic, staff wellbeing has been considered as the most critical issue in organizational change management. Following an overview of the relevant literature and recent official statistics, this essay aims to (i) address the major considerations and challenges in light of the transformation and re-design of the mode of work in the new normal and (ii) inform practical decisions for overall staff wellbeing under post-pandemic work-from-home (WFH) conditions with recommendations. For the sake of both staff healthiness and safety, as well as organizational competitiveness, senior management should take reasonable steps to enhance occupational safety in their WFH policy in line with practical recommendations on five areas, namely, (i) ergonomics, (ii) stress and anxiety management, (iii) workplace boundaries, (iv) work–family conflicts, and (v) other factors regarding a negative work atmosphere (e.g., loneliness attack, burnout, and workplace violence) particularly on virtual platforms. With the suggested evidence-based practices on WFH initiatives, senior management could make a difference in optimizing the overall workplace wellbeing of staff after the pandemic.

Keywords: COVID-19; teleworking; personnel management; occupational health; organizational innovation

check for updates

Citation: Cheung, V.K.L. Practical Considerations of Workplace Wellbeing Management under Post-Pandemic Work-from-Home Conditions. *Int. J. Environ. Res. Public Health* **2024**, *21*, 924. https://doi.org/10.3390/ijerph21070924

Academic Editor: Masahito Fushimi

Received: 27 May 2024
Revised: 10 July 2024
Accepted: 11 July 2024
Published: 15 July 2024

1. Introduction

1.1. Definition and Concept of Work from Home

The concept of work from home (WFH) originated from Nilles [1], using the terminologies "telework" and "telecommuting" to describe how employees perform duties in flexible workplaces wherever technology supports the communication and completion of tasks. WFH, interchangeable with remote work, e-working, home or virtual offices, became popular during the COVID-19 pandemic [2]. According to the Labor Force Survey, the proportion of WFH employees in the UK increased from 6% in 2017 to over 80% in the spring of 2020 [3]. In the United States, the proportion of WFH employees increased from 10% to 35% (or over 60% full-time employees), with over 70% of employees reporting a self-perceived work effectiveness [4].

Whether or not WFH is suitable as a mode of work for densely populated cities remains inconclusive [4]. Researchers found that the majority (80%) of employees in Hong Kong prefer WFH (either mixed or hybrid) to traditional office work as they appreciate more rest time (72%), lower work stress (>60%), less mental exhaustion (>80%), and flexible and compressed working hours (around 80%) [5]. With these appealing preliminary data, the following paragraphs will address the considerations of workplace wellbeing with theory and evidence in line with guidelines from BPS.

1.2. Psychological Models of WFH Arrangements

An analytic framework has been established to inform structural considerations for a WFH arrangement and systemic effect on a variety of performance indicators during the pandemic [6]. In this model, "organizational" and "individual and family" factors determine outcomes on "occupation domain" (e.g., productivity, job satisfaction, flexibility, work engagement, and presenteeism/absenteeism) and "family domain" (e.g., work–life balance, life/family satisfaction, personal and family health conditions) (Figure 1).

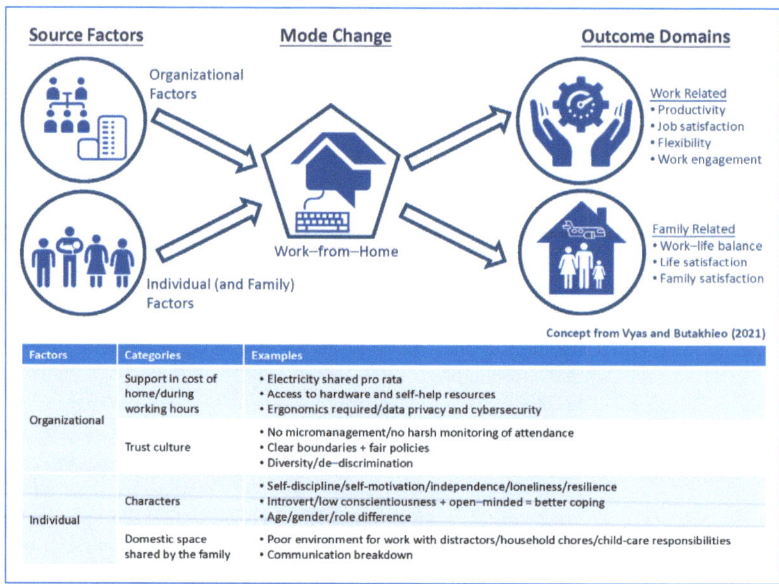

Factors	Categories	Examples
Organizational	Support in cost of home/during working hours	• Electricity shared pro rata • Access to hardware and self-help resources • Ergonomics required/data privacy and cybersecurity
	Trust culture	• No micromanagement/no harsh monitoring of attendance • Clear boundaries + fair policies • Diversity/de–discrimination
Individual	Characters	• Self-discipline/self-motivation/independence/loneliness/resilience • Introvert/low conscientiousness + open–minded = better coping • Age/gender/role difference
	Domestic space shared by the family	• Poor environment for work with distractors/household chores/child-care responsibilities • Communication breakdown

Figure 1. Analytic Framework of Work from Home during COVID-19 Pandemic.

With reference to guides on COVID-specific anxiety and distress in the workplace [7] and that on WFH conditions [8], the "SHARE" model has been disseminated as a psychologically informed approach for considerations towards a healthy and sustainable environment for WFH arrangements during the pandemic, namely, (i) safe homeworking, (ii) help yourself and others, (iii) adapt to change, (iv) relieve the pressure, and (v) evaluate (Table 1).

Table 1. SHARE Model from the British Psychological Society (BPS) [7,8].

Safe homeworking	• Duty of care for staff safety • Practical guide on roster, data privacy regulations, and equipment support • Fairness/communication channels • Ergonomics
Help yourself and others	• Communicating and meeting needs $\geq$ set realistic targets • Communication and check-ins/time-off regulations • Technical/resilience skills • Support hidden costs
Adapt to change	• Diverse workplace situations and adjusting to the new normalcy for Diversity, Equity, and Inclusion (DEI) • Personality/Individual differences • Risk of cyberbullying/workplace violence • Role conflict/work–life balance $\geq$ boundary setting
Relieve the pressure	• Helping to adapt and cope with pressure from work and family
Evaluate	• Reviewing the conditions regularly to ensure sustainable success (e.g., workspace and work format design/redesign)

Due to the complicated relationship among multi-faceted elements, these analytic frameworks and guides help to structuralize key components for managing workplace wellbeing: occupation safety and health, stress and anxiety at work, boundaries, family conflict, and workplace conflict and violence.

2. Major Considerations

2.1. Occupational Safety and Health Consideration—Ergonomics

According to BPS [7,8], employers should have accountability for the workplace wellbeing of employees. Duty of care must include a risk assessment in the workplace, where ergonomics or human factors are key to workplace safety and health. The first practical challenge is ergonomically caused physical injury.

The Institute of Employment Studies [9] was commissioned during the COVID-19 lockdown to explore whether WFH arrangements would be good for employee wellbeing. Despite the coexisting effect of stress and anxiety induced by the lockdown or individual difficulties, non-standardized workplace settings triggered more complaints of musculoskeletal discomfort and unsatisfactory work–life balances from over half of the respondents. Gerding et al. [10] explored how ergonomics in home offices influenced the physical health of university staff members ($n = 843$). They found that compared with university office environments, participants had worked under suboptimal conditions (45% worked on a seat without adjustable armrests; 85% worked with a laptop instead of a desktop with a monitor and mouse). Despite reports including subjective feelings and having no track record for the actual onset date of physical discomfort, 40% of participants reported moderate-to-severe discomfort (e.g., severe low-back pain and moderate discomfort from head to shoulders) and worrying about developing musculoskeletal disorder in the long run.

Given the basic requirements of a workplace setup (e.g., the acquisition of technical skills, a computer, and an IT network, access to the necessary resources, and a connection to virtual teams), a non-standardized work environment, such as a sub-vented home with other co-tenants or one with too many distractors nearby, would pose a threat of cybersecurity or data privacy issues to employees and the organization. Managers should review the applicability of WFH arrangements with employees (e.g., the assessment of "Display Screen Equipment" covering all equipment during computer use), including the employees' home conditions for occupational safety and health (OSH) purposes, enabling them to access the necessary technological support with cybersecurity devices and financially with additional devices (e.g., a work-specific computer and software and a chair with an armrest).

2.2. Stress and Anxiety at Work

Compared with family and other health-related issues, work has become a main source of stress in the general labor population [11]. The effects of chronic workplace stress regarding the wellbeing of staff, including excessive alcohol consumption, cardiovascular accidents, burnout, musculoskeletal disorders, overweight or obesity scores, eating and mood disorders, and mortality, have led to increased public awareness. This concerns its immense societal impacts on the overall productivity and costs of health insurance, as well as the healthcare burden on the global economy, which is over a trillion a year [12,13].

2.2.1. Would WFH Conditions Lead to Positive Behavioral Change in Health Outcomes?

Researchers from Sweden explored the relationship of health conditions and behavioral change, as well as the distribution of time for sleep, work, leisure activity, and physical activity, by interviewing 27 full-time workers under WFH conditions [14]. Compared with the mode of office work, WFH arrangements resulted in an increased duration of sleep, largely saved from the non-office-hour work time (reduced OT). Quantity does not guarantee quality. At the other extreme, oversleeping would be detrimental to one's physical and mental health [15]. Models addressing the degree to which sleeping time explains wellbeing outcomes and its association with physical activity and quality of life would be suggested for further studies during the post-pandemic phase.

2.2.2. Do Personality Types Matter in WFH Conditions?

It is expected that people with a higher mood stability (or lower neuroticism) would cope better in WFH conditions because they were less likely pre-occupied with negative thoughts by working alone, which in turn can cause a psychological burden [11]. A common predictor for "excellent staff" in the workplace, conscientiousness, may be an interesting character under WFH conditions [16]. Based on about 200 sets of field data, Abbas and Raja [11] found that hindrance stressors were associated with psychological strains and turnover. In the face of the ever-evolving working environment during the pandemic, employees at a low end of conscientiousness (especially those with high openness) had lower turnover rates as high achievers because they tended to cope better with hindrance stressors (e.g., role ambiguity and overload) than their highly conscientious counterparts who stuck to deep-rooted mindsets in fixed and organized business landscapes [11]. The reasons for the conscientious employees leaving the organization were burnout, followed by overworking (with it being difficult to "switch-off") and maladaptation as regards changes.

2.2.3. Sense of Control and Job Satisfaction

In addition to the adverse health outcomes to get rid of, managers should take into account the positive aspects of workplace wellbeing, such as job satisfaction. Deci et al. [17] claimed that in order to foster job satisfaction, three senses, namely, autonomy, relatedness, and competence, are facilitators. Goh et al. [18] found that workplace wellbeing (e.g., physical and psychological health conditions and morbidity and mortality rates) was contributive to job control as regards the working pattern (e.g., flexi working hours and boundaries), self-autonomy (e.g., prioritization and workload), and power for decision making (communication, engagement, and teamwork), where the perceived fairness (with a constructive health policy in place) in the decision-making process and the interpersonal treatment throughout the course of the WFH period change the magnitude between them. A lack of control would likely consume more mental resources, decrease self-efficacy, and put employees at risk of burnout displayed by emotional exhaustion, depersonalization, and loss of personal accomplishment [19]. When job stability and flexibility were maintained in home working environments, the sense of security (e.g., daily engagement and finance source) and sense of control in each employee's own pace of work may potentially trigger job satisfaction, higher productivity, less stress-induced anxiety, and conflict with family and peers [19–21]. The next paragraph will cover the control of boundary and work–family balance.

2.3. Workplace Boundaries

The negative influence of WFH arrangements has resulted in longer working hours, an isolation-induced sense of loneliness, and an invasion of "time and space" for families, as well as greater family influence on work [22]. Compared with home workers, office workers would have a relatively clear boundary: When I am off-duty, I leave the office physically. In the sense of "process of psychological detachment" or "switch-off", physical attendance in an office has a symbolic meaning of "I am at work", while leaving the office would be for "pastime". WFH arrangements are actually breaking ground to make a physical and psychological separation between the work and office indifferent. Since the work–life boundary was blurred, Grant et al. [21] found that, under these conditions, employees tended to overwork during the pandemic, in particular those living with distractors (e.g., children under care) and high disengagement with colleagues [19,23].

To what extent and dimension should employers be accountable for employees' wellbeing out of the office, in terms of fixed working hours, flexi-hours on an individual basis, or workload based on task output? Irrespectively, a clear boundary policy of the working hours and workload would be appreciated. Managers should also be aware of how excessive work demands affect not only the health condition of employees but also their relationship with family members.

2.4. Work–Family Conflicts

WFH arrangements would definitely put employees on a "stress test" under mixed conditions with both work–family interferences potentially being conflict-evoking. The classic scarcity hypothesis [24] indicated that personal resources, such as time, energy, and money, are limited and consumed under a zero-sum game: the more you invest certain resources into work, the less you can maintain a good relationship with your family. On the contrary, Solis [25] argued that, given fixed working schedules and individual spaces in home offices, time originally spent on traveling could be spent on building positive relationships with family members. Some scholars have examined relevant theoretical concepts by exploring how pressure from work triggers or intensifies inter-role conflicts with family members [26,27], categorizing work–family conflicts into different forms, namely, strain-based (negative emotions or pressure from one role affecting one's fulfilment of another role), time-based (not meeting the obligations of the two roles due to time collisions) and behavior-based (the same patterns or characteristics not being compatible in two roles) across different magnitudes, such as family interference from work or work interference from family.

According to Giovanis et al. [2,28], common predictors in work–family conflicts include:

(i). Sex: Female WFH employees reported a positive life satisfaction and family relationship as the allocation of household chores and family responsibility was shared with their partners. Correspondingly, male employees' slight decrease in job performance and productivity suggested a greater work influence from family than their partners.

(ii). Parental status: A positive association between the number of children the employee had and "work interference from family" were found. Married parents, especially those with a clear role delineation as breadwinner and houseworker, reported less inter-role conflicts than single parents, regardless of age and sex.

(iii). Other situational factors: Role stressors (role conflict, ambiguity, demands, and overload) showed a negative association with the role's involvement in work–family conflicts, which implied a detachment from the role of work.

(iv). Which one has a greater impact on work–family outcomes, "family interference from work" or "work interference from family"? Regarding the indicators of satisfaction, job satisfaction is highly associated with "work interference from family", while personal satisfaction (e.g., life or marriage) is highly associated with "family interference from work".

(v). Psychological and physical strains are important indicators that link to the work–family relationship. For instance, burnout and alcoholism are more affected by a high level of work–family conflicts, particularly "work interference from family", while a high-fat diet and physical inactivity were associated with "family interference from work". Furthermore, work–family conflicts are attributive to poor sleep quality, amount of time spent asleep and nightmare frequency.

Irrespective of the direction of the interference on work–family conflicts, managers should provide the necessary measures to relieve pressure and evaluate its effects on employee wellbeing [7,8]. As the last resort, a psychological support service as well as an alternative WFH environment may be considered.

2.5. Workplace Conflict and Violence in WFH Arrangements

WFH arrangements can be a double-edged sword to workplace harmony and destructive behavior in an organization. Arvola et al. [22] were optimistic about the positive influence of WFH conditions on autonomy (reducing direct confrontation in supervision), accountability, and professionalism, which would be extremely favorable to younger employees. However, Wech et al. [29] conducted a series of qualitative studies to prove that, except for the lack of communication, post-pandemic anxiety followed by excessive workloads and the change to WFH mode and process would increase cyberbullying and peer conflicts with passive-aggressive behavior as well as malfunctioned social competition.

Virtual platforms for WFH environments may not be easily adapted to by certain groups of employees, such as senior staff, those prone to digital fatigue or anxiety, and those with a physical disability. Such sudden changes in the work format without preparedness may trigger job insecurity or dissatisfaction or, even worse, exacerbate virtual team conflicts as well as workplace violence in digital form, such as online abuse/cyberbullying or spread of hate speech on inappropriate content (e.g., derogatory or obscene) [29].

Age discrimination following poor adaption to the specific technology required for WFH conditions is generally unprotected by law in most countries. Regarding seniority and specialty, additional support in advanced intercommunication and technology learning, good communication with complete information, and a decision-making process with the virtual teams could reduce the turnover rate of senior specialists and ensure productivity [30].

Affected by quarantine or lockdown under the pandemic, uncertainty and fear of being infected, anger for the restriction of outdoor activity, and the hatred of certain ethnic or minority groups have grown exponentially [2]. Cyberbullying, such as hostile and hate speech, and pointless accusations about certain racial groups spreading the virus may be the tip of the iceberg, particularly for those working with virtual teams comprised of members with diverse cultural background. A conflict resolution workshop and stress and emotion management, as well as a resilience workshop with a self-help toolkit, could be incorporated into any existing core competence training for internal staff on the virtual platform in order to dismantle discriminative or microaggressive behaviors under WFH conditions, such as gaslighting [30].

3. WFH-Specific Challenges: Social Contact—Loneliness and Wellbeing

The classic social comparison theory from Festinger [31] and the social learning theory from Bandura [32] asserted that individuals tended to reduce uncertainty by comparing the perception of others and themselves on certain statuses (e.g., sense of achievement) with internal feedback as an "objective benchmark" and be motivated to replicate behavior by observing peers with reciprocity and mutual goals. Beyond simply surviving as a human being, work can evoke a positive experience with the social contact of friendship, sense of identity, achievement on task completion, purpose or meaning of life fulfilment, and personal growth with resilience and character strengths when tackling a crisis [33,34]. Without face-to-face interaction under WFH conditions, employees would be attacked by loneliness in comparison with those having frequent direct interaction in the workplace.

As a result, the sense of loneliness or lack of socialization under isolation without timely feedback and mutual support from peers reduced self-efficacy and increased anxiety levels, which would result in overworking and burnout [2,35]. When people were urged to prove their value at work, they started to lose the work–life boundary in order to increase productivity, which might significantly reduce private time for their beloved ones (low social support) and taking a rest (low sleeping hours).

With 50% burnout rate in the global workforce [13], the real proportions and numbers may be underestimated because seriously burned-out employees would have remained unresponsive or quit their current full-time positions. From another angle, Collins et al. [36] argued that WFH arrangements provided staff who disliked their office environment with self-control on building positive relationships with those they want and avoiding conflicts with unfavorable peers. Nevertheless, the downsides of WFH conditions as regards team collaboration and communication with creativity and diversity remained non-demystified.

4. Power of Communication and Standardized Operation Guide

Ethically speaking, WFH arrangements should not be used by an organization for the sake of oppression. To strengthen WFH wellbeing as well as the productivity of the organization, communication with valid WFH policies, guidelines, and approaches is critical. McKinney and Company [13] indicated that WFH wellbeing (feeling supported and included) and productivity would show a two-to-five times increase when clear communication on relevant visions and policies has been achieved. Of over 5000 full-time employees, two-thirds were concerned about the existing communication breakdown in WFH approaches. When management supported in setting the boundary policy and team communication, as well as when the training of digitalization (on both technical and soft-skill-handling workload) was given, behavioral health and mental health conditions (e.g., depression, anxiety, and stress) of WFH employees were largely improved by 70% [12,14]. With the limitation of the large sample variation for synthesized findings, the inter-relatedness of the mentioned factors and the degree to which they could explain the overall wellbeing that followed WFH arrangements remained unknown.

5. Evidence-Backed Recommendation on Occupation Safety in WFH Arrangements

Actionable insights regarding WFH arrangements have been consolidated with practical recommendations applicable to occupational safety management.

5.1. Considerations in Ergonomics

To reduce the risk of acute physical injury (e.g., fracture from fall, neck sprain, etc.) and chronic health hazards (e.g., low back pain, musculoskeletal disorder, etc.), performing a preliminary assessment in determining the eligibility of the occupational safety of potential WFH environment is pivotal.

- In cases where the assessed environment was either suboptimal or impossible to improve with supplementary equipment offered by the company, a distanced-but-standardized workplace setting (such as a rental office or a library conference room with seats with adjustable armrests and an office notebook) would be much more suitable as opposed to a non-standardized workplace setting.
- Even if the assessed environment was considered largely eligible, managers are accountable for providing necessary operational, technical, and technological support to enable workers to execute their job duties under physically and psychologically safe conditions. They are also accountable for providing equipment and an environment assessment on a regular basis.

5.2. Stress and Anxiety at Work

To mitigate occupational stress- and anxiety-induced health hazards (e.g., sleep disturbance, being overweight, cardiac illness, alcoholism, etc.), reinforcing measures on healthy lifestyle modification, personality, and job satisfaction should be taken into account.

- With psychoeducation workshops implemented by healthcare professionals and psychologists who underwent qualified training in motivational interviewing techniques, it is anticipated that staff members under WFH arrangements could cope better with occupational stress and anxiety following changes in healthy lifestyle patterns, such as regular physical activity, cessation of excessive smoking, low alcohol consumption, and non-disruptive work and sleep quality with minimal doom-scrolling or midnight video conferences.
- Personality matters in workplace adaptation. Counterintuitively, staff members with a high conscientiousness might require extra guidance on WFH arrangements. The ever-changing nature of role ambiguity and overloading under unclear work boundaries took a massive toll on their physical and psychological health, leading to high burnout and turnover rates. Recapitulating the "switch-off" policy for working hours and active coaching on effective stress-coping techniques as well as time and workload management would be beneficial in reducing the burnout and turnover rates of valuable staff in the WFH mode.
- For positive aspects, ways to enhance job control and satisfaction guarantee the occupational wellbeing of staff and the sustainable development of the organization. Under a fair system of WFH conditions, managers could set the standard requirements (e.g., number of working hours, case load, formal meeting arrangements, recess intervals, etc.) and then allow flexibility for staff to maintain the sense of control by

prioritizing their tasks, maintaining job satisfaction by being respectfully engaged in the decision-making process with optimal psychological safety.

5.3. Workplace Boundaries

Uncertainty of working boundaries, including time and space with family members and the sense of loneliness under a WFH status, must be handled with care.

- It is necessary for senior management to maintain a balance between setting boundaries (or clear switch-on and switch-off times) and providing ready support from supervisors and colleagues via a chosen electronic platform during official working hours.
- Asides from having a policy of standard working hours and workload in place, extra attention might be put towards the "overwork tendency" of those who have children to take care of in distanced workplaces and low engagement with peers. Supervisors may provide members in this target group with proactive mentoring on the progress of work and short-but-regular video conferences to strengthen the connectivity among team members on a weekly basis at minimum.

5.4. Work–Family Conflicts and Workplace Violence

It is essential for an organization to clearly define the sources of stress from the workplace and enrich their understanding of staff at all levels, informing staff about the related stress-reduction measures at primary, secondary, and tertiary intervention levels for stress management from a public health perspective.

- Special care should be given to those who have a higher proclivity to work–family conflicts (and lower job performance) under WFH conditions: (i) male, (ii) more children, (iii) unclear family role delineation if married, (iv) high role stressors at work, and (v) unhealthy lifestyle (e.g., poor sleep, alcoholism, physical inactivity, etc.).
- Regarding strain-based stress, managers might help staff to reduce role conflicts induced within the workplace or from their families by setting clear role-specific performance indicators in several phases. This could help staff to regain a level of control over demanding tasks towards feasible goals in the short term and long term.
- Regarding time-based stress, managers could reduce time collision by setting clear work boundaries and allowing flexibility for staff to complete tasks under reasonable timeframes based on personal schedules. Job redesign for specification and timeline may be required if regular reviews indicate that certain tasks are behind schedule owing to severe understaffing, high turnover rates, or absenteeism.
- Regarding behavior-based stress, managers may consider referring staff to a secondary intervention (e.g., relaxation, mindfulness or meditation group, psychosocial intervention training, or brief cognitive behavioral therapeutic sessions) managed by healthcare professionals. Individual fitness and an early stress-case-management program would be helpful in preventing the exacerbation of mild, unhealthy stress-coping patterns. In cases involving behavioral issues, whether highly complex or personal, immediate action regarding a preliminary workplace assessment and referral to a tertiary intervention (e.g., individual counseling or psychotherapy) should be taken in a timely manner. In addition to a psychological support service, an alternative WFH environment as opposed to a home environment after re-assessing the eligibility of the WFH arrangement may be arranged as the last resort.
- WFH-specific training with a live demonstration on how to activate off-site access to the database, work schedule, and communication platform and an enquiry hotline for technical, operational, and emotional support within office hours would be helpful in reducing any communication breakdown (e.g., excessive anxiety, maladaptive mode of work, and peer conflicts, etc.) and sense of loneliness. Establishing a robust WFH-specific workplace violence policy framework could build on the momentum of the culture change to dismantle cyberbullying at an organization-wide level.

With these suggested evidence-based practices for WFH initiatives, senior management could work together with all of the stakeholders to optimize sustainable growth in the wellbeing of their employees as well as in competitiveness of their organization after the pandemic.

Funding: This study received no specific grant from any funding agency in the public, commercial, or not-for-profit sectors in affiliation with the University of Leicester.

Institutional Review Board Statement: Not applicable.

Informed Consent Statement: Not applicable.

Data Availability Statement: Not applicable.

Conflicts of Interest: The author declares no conflicts of interest.

References

1. Nilles, J.M. Telework: Enabling distributed organizations: Implications for IT managers. *Inf. Syst. Manag.* **1997**, *14*, 7–14. [CrossRef]
2. Rudolph, C.W.; Allan, B.; Clark, M.; Hertel, G.; Hirschi, A.; Kunze, F.; Shockley, K.; Schoss, M.; Sonnentag, S.; Zacher, H. Pandemics: Implications for research and practice in industrial and organizational psychology. *Ind. Organ. Psychol. Perspect. Sci. Pract.* **2020**, *14*, 1–35. [CrossRef]
3. Office for Nation Statistics (ONS); Cameron, A. Coronavirus and Homeworking in the UK—Office for National Statistics. Gov.uk. Published 7 July 2020. Available online: https://www.ons.gov.uk/employmentandlabourmarket/peopleinwork/employmentandemployeetypes/bulletins/coronavirusandhomeworkingintheuk/april2020 (accessed on 3 November 2022).
4. Bick, A.; Blandin, A.; Mertens, K. Work from home before and after the COVID-19 outbreak. *SSRN Electron. J.* **2021**. Available online: https://doi.org/10.2139/ssrn.3786142 (accessed on 3 November 2022).
5. Wong, H.K.A.; Cheung, J. Survey Findings on Working from Home under COVID-19. Lingnan University. Available online: https://commons.ln.edu.hk/covid-survey/1/ (accessed on 3 November 2022).
6. Vyas, L.; Butakhieo, N. The impact of working from home during COVID-19 on work and life domains: An exploratory study on Hong Kong. *Policy Des. Pract.* **2020**, *4*, 59–76. [CrossRef]
7. Waites, B.; Breslin, G.; Thomson, L. COVID-Related Anxiety and Distress in the Workplace: A Guide for Employers and Employees. Ulster.ac.uk. [BPS2020a]. Available online: https://pure.ulster.ac.uk/ws/files/85822891/Covid_related_anxiety_and_stress_in_the_workplace_Waites_et_al_2020.pdf (accessed on 3 November 2022).
8. Kinman, G.; Breslin, G.; Thomson, L.; Mackey, L. Working from Home: Healthy Sustainable Working during the COVID-19 Pandemic and beyond. Ulster.ac.uk. [BPS2020b]. Available online: https://pure.ulster.ac.uk/ws/files/79506622/Working_from_home_Kinman_et_al_2020.pdf (accessed on 3 November 2022).
9. Institute of Employment Studies (IES). Homeworker Wellbeing Survey. Available online: https://www.employment-studies.co.uk/sites/default/files/resources/summarypdfs/IES%20Homeworker%20Wellbeing%20Survey%20Headlines%20-%20Interim%20Findings.pdf (accessed on 1 November 2021).
10. Gerding, T.; Syck, M.; Daniel, D.; Naylor, J.; Kotowski, S.E.; Gillespie, G.L.; Freeman, A.M.; Huston, T.R.; Davis, K.G. An assessment of ergonomic issues in the home offices of university employees sent home due to the COVID-19 pandemic. *Work* **2021**, *68*, 981–992. [CrossRef]
11. Abbas, M.; Raja, U. Challenge-hindrance stressors and job outcomes: The moderating role of conscientiousness. *J. Bus. Psychol.* **2019**, *34*, 189–201. [CrossRef]
12. Evanoff, B.A.; Strickland, J.R.; Dale, A.M.; Hayibor, L.; Page, E.; Duncan, J.G.; Kannampallil, T.; Gray, D.L. Work-related and personal factors associated with mental well-being during the COVID-19 response: Survey of health care and other workers. *J. Med. Internet Res.* **2020**, *22*, e21366. [CrossRef]
13. McKinney & Company; Alexander, A.; De Smet, A.; Ravid, D. What Employees Are Saying about the Future of Remote Work. Mckinsey.com. Published 1 April 2021. Available online: https://www.mckinsey.com/business-functions/people-and-organizational-performance/our-insights/what-employees-are-saying-about-the-future-of-remote-work (accessed on 3 November 2022).
14. Hallman, D.M.; Januario, L.B.; Mathiassen, S.E.; Heiden, M.; Svensson, S.; Bergström, G. Working from home during the COVID-19 outbreak in Sweden: Effects on 24-h time-use in office workers. *BMC Public Health* **2021**, *21*, 528. [CrossRef]
15. Stamatakis, K.A.; Punjabi, N.M. Long sleep duration: A risk to health or a marker of risk? *Sleep. Med. Rev.* **2007**, *11*, 337–339. [CrossRef]
16. Dudley, N.M.; Orvis, K.A.; Lebiecki, J.E.; Cortina, J.M. A meta-analytic investigation of conscientiousness in the prediction of job performance: Examining the intercorrelations and the incremental validity of narrow traits. *J. Appl. Psychol.* **2006**, *91*, 40–57. [CrossRef]

17. Deci, E.L.; Ryan, R.M.; Gagné, M.; Leone, D.R.; Usunov, J.; Kornazheva, B.P. Need satisfaction, motivation, and well-being in the work organizations of a former eastern bloc country: A cross-cultural study of self-determination. *Pers. Soc. Psychol. Bull.* **2001**, *27*, 930–942. [CrossRef]
18. Goh, J.; Pfeffer, J.; Zenios, S.A.; Rajpal, S. Workplace stressors & health outcomes: Health policy for the workplace. *Behav. Sci. Policy* **2015**, *1*, 43–52. [CrossRef]
19. Kazekami, S. Mechanisms to improve labor productivity by performing telework. *Telecommun. Policy* **2020**, *44*, 101868. [CrossRef]
20. Arntz, M.; Sarra, B.Y.; Berlingieri, F. Working from home: Heterogeneous effects on hours worked and wages. *SSRN Electron. J.* **2019**. Available online: https://doi.org/10.2139/ssrn.3383408 (accessed on 3 November 2022).
21. Grant, C.A.; Wallace, L.M.; Spurgeon, P.C.; Tramontano, C.; Charalampous, M. Construction and initial validation of the E-Work Life Scale to measure remote e-working. *Empl. Relat.* **2019**, *41*, 16–33. [CrossRef]
22. Arvola, R.; Tint, P.; Kristjuhan, Ü.; Siirak, V. Impact of telework on the perceived work environment of older workers. *Sci. Ann. Econ. Bus.* **2017**, *64*, 199–214. [CrossRef]
23. Eddleston, K.A.; Mulki, J. Toward understanding remote workers' management of work-family boundaries: The complexity of workplace embeddedness. *Group Organ. Manag.* **2017**, *42*, 346–387. [CrossRef]
24. Goode, W.J. A theory of role strain. *Am. Sociol. Rev.* **1960**, *25*, 483. [CrossRef]
25. Solís, M.S. Telework: Conditions that have a positive and negative impact on the work-family conflict. *Acad. Rev. Latinoam. Adm.* **2016**, *29*, 435–449. [CrossRef]
26. Greenhaus, J.H.; Beutell, N.J. Sources of conflict between work and family roles. *Acad. Manag. Rev.* **1985**, *10*, 76. [CrossRef]
27. Kahn, R.L.; Wolfe, D.M.; Quinn, R.P.; Snoek, J.D.; Rosenthal, R.A. *Organizational Stress: Studies in Role Conflict and Ambiguity*; John Wiley & Sons: Hoboken, NJ, USA, 1964.
28. Giovanis, E. Are women happier when their spouse is teleworker? *J. Happiness Stud.* **2018**, *19*, 719–754. [CrossRef]
29. Wech, B.A.; Howard, J.; Autrey, P. Workplace bullying model: A qualitative study on bullying in hospitals. *Empl. Responsib. Rights J.* **2020**, *32*, 73–96. [CrossRef]
30. Melo, P.C.; de Abreu e Silva, J. Home telework and household commuting patterns in Great Britain. *Transp. Res. Part A Policy Pract.* **2017**, *103*, 1–24. [CrossRef]
31. Festinger, L. A theory of social comparison processes. *Hum. Relat.* **1954**, *7*, 117–140. [CrossRef]
32. Bandura, A. Self-efficacy: Toward a unifying theory of behavioral change. *Psychol. Rev.* **1977**, *84*, 191–215. [CrossRef] [PubMed]
33. Paul, K.I.; Batinic, B. The need for work: Jahoda's latent functions of employment in a representative sample of the German population: The need for work: Latent functions of employment. *J. Organ. Behav.* **2010**, *31*, 45–64. [CrossRef]
34. Seligman, M.E.P.; Csikszentmihalyi, M. Positive Psychology: An Introduction. In *Flow and the Foundations of Positive Psychology*; Springer: Dordrecht, The Netherlands, 2014; pp. 279–298.
35. Biron, M.; van Veldhoven, M. When control becomes a liability rather than an asset: Comparing home and office days among part-time teleworkers: Within-individual Study on Part-time Telework. *J. Organ. Behav.* **2016**, *37*, 1317–1337. [CrossRef]
36. Collins, A.M.; Hislop, D.; Cartwright, S. Social support in the workplace between teleworkers, office-based colleagues and supervisors. *New Technol. Work. Employ.* **2016**, *31*, 161–175. [CrossRef]

International Journal of
Environmental Research and Public Health

Article

Adapting the Stress First Aid Model for Frontline Healthcare Workers during COVID-19

Mayer H. Bellehsen [1,2,3], Haley M. Cook [1,4,*], Pooja Shaam [1,4], Daniella Burns [1,3], Peter D'Amico [2], Arielle Goldberg [1,3], Mary Beth McManus [3], Manish Sapra [2,3], Lily Thomas [5], Annmarie Wacha-Montes [1,3], George Zenzerovich [1,3], Patricia Watson [6], Richard J. Westphal [7] and Rebecca M. Schwartz [1,2,4]

[1] Center for Traumatic Stress, Resilience and Recovery at Northwell Health, Great Neck, NY 11021, USA; mbellehsen@northwell.edu (M.H.B.); pshaam@northwell.edu (P.S.); dburns1@northwell.edu (D.B.); agoldberg9@northwell.edu (A.G.); awachamontes@northwell.edu (A.W.-M.); gzenzerovich@northwell.edu (G.Z.); rschwartz3@northwell.edu (R.M.S.)

[2] Department of Psychiatry, Zucker Hillside Hospital at Northwell Health, Glen Oaks, NY 11004, USA; pdamico@northwell.edu (P.D.); msapra@northwell.edu (M.S.)

[3] Behavioral Health Service Line, Northwell Health, New York, NY 10022, USA; mmcmanus@northwell.edu

[4] Department of Occupational Medicine, Epidemiology and Prevention, Donald and Barbara Zucker School of Medicine at Hofstra/Northwell, Great Neck, NY 11021, USA

[5] Institute for Nursing, Northwell Health, New Hyde Park, NY 11042, USA; lthomas@northwell.edu

[6] National Center for PTSD, White River Junction, VT 05009, USA; patricia.j.watson@dartmouth.edu

[7] School of Nursing, University of Virginia, Charlottesville, VA 22903, USA; richard.westphal@virginia.edu

* Correspondence: hcook1@northwell.edu; Tel.: +1-(516)-465-3177 (ext. 8383)

Abstract: The coronavirus pandemic has generated and continues to create unprecedented demands on our healthcare systems. Healthcare workers (HCWs) face physical and psychological stresses caring for critically ill patients, including experiencing anxiety, depression, and posttraumatic stress symptoms. Nurses and nursing staff disproportionately experienced COVID-19-related psychological distress due to their vital role in infection mitigation and direct patient care. Therefore, there is a critical need to understand the short- and long-term impact of COVID-19 stress exposures on nursing staff wellbeing and to assess the impact of wellbeing programs aimed at supporting HCWs. To that end, the current study aims to evaluate an evidence-informed peer support stress reduction model, Stress First Aid (SFA), implemented across units within a psychiatric hospital in the New York City area during the pandemic. To examine the effectiveness of SFA, we measured stress, burnout, coping self-efficacy, resilience, and workplace support through self-report surveys completed by nurses and nursing staff over twelve months. The implementation of SFA across units has the potential to provide the workplace-level and individual-level skills necessary to reduce stress and promote resilience, which can be utilized and applied during waves of respiratory illness acuity or any other healthcare-related stressors among this population.

Keywords: Stress First Aid; nurses; COVID-19; resilience; burnout; stress

Citation: Bellehsen, M.H.; Cook, H.M.; Shaam, P.; Burns, D.; D'Amico, P.; Goldberg, A.; McManus, M.B.; Sapra, M.; Thomas, L.; Wacha-Montes, A.; et al. Adapting the Stress First Aid Model for Frontline Healthcare Workers during COVID-19. *Int. J. Environ. Res. Public Health* **2024**, *21*, 171. https://doi.org/10.3390/ijerph21020171

Academic Editor: Masahito Fushimi

Received: 5 December 2023
Revised: 22 January 2024
Accepted: 30 January 2024
Published: 1 February 2024

1. Introduction

The COVID-19 pandemic placed an unprecedented burden on healthcare workers (HCWs). It also highlighted pre-existing stressors that exacerbate challenges within hospital systems across the country [1,2]. HCWs experience high levels of stress due to the nature of their job responsibilities, including caring for critically ill and injured patients, patient deaths, increased workloads, and long hours [3]. These occupational stressors are often exacerbated during periods of higher healthcare demand, such as during a disaster or public health emergency [4,5]. Occupationally, HCWs are particularly vulnerable to workplace stressors, conferring increased risk for posttraumatic stress reactions [5,6]. Indeed, an analysis of the 2008 American Community Survey and National Death Index records through 2019 found that healthcare workers were at increased risk of suicide compared with non-healthcare workers, specifically registered nurses, healthcare support workers, and health technicians [7]. National and international studies of HCWs' occupational and health outcomes during the COVID-19 pandemic have shown increased rates of psychopathology including anxiety, depression, and posttraumatic stress; behavioral changes including sleep disturbance, relationship difficulties, and substance use; and increased rates of burnout, compassion fatigue, and job dissatisfaction [1,2,6,8–10].

Approximately two years after the first wave of the pandemic, the US Bureau of Labor Statistics estimated that there were over three million employed registered nurses, the largest healthcare occupation in the US, of which nearly 88% identify as female [10,11]. Nursing staff report increased rates of moral injury stemming from treatment decisions based on limited resources, increases in stress from the volume and intensity of their work, and a new sense of vulnerability and fear for their safety and the safety of their loved ones [12,13]. Similar outcomes were found among nurses across triage and non-triage hospitals in Egypt during the pandemic, wherein occupational stressors including high workload, exposure to death, personal fears, and stigma were associated with higher stress, decreased job satisfaction, and increased intent to leave their current position [14]. In a longitudinal study of Italian healthcare workers between May 2020 and July 2021, the authors found subclinical levels of psychiatric symptoms, including stress, depression, state anger, and emotional exhaustion, across the three time points measured in the study, hypothesized to indicate a higher baseline level of resilience among HCWs [15]. Related measures of emotional and occupational wellbeing, however, were found to be negatively impacted in a another sample of Italian healthcare workers, during (July 2021) and post (July 2023) pandemic [16]. Specifically, while rates of job burnout and symptoms of depression remained high between time points, rates of compassion fatigue increased following the pandemic [16]. Findings may be suggestive of negative ongoing impacts to emotional and occupational wellbeing experienced by healthcare workers. The collective impact of trauma from COVID-19, as well as subsequent responses to the pandemic at the provider level, contribute to the emotional toll experienced in the wake of COVID-19 on healthcare providers [1,2]. With an understanding of the unique risks experienced by HCWs generally and those experienced explicitly by nurses during the COVID-19 pandemic, targeted interventions are needed to address provider wellbeing.

1.1. Peer Support Intervention

The relationship between occupational stress and adverse psychological outcomes is well established [17–20]. Peer support programs have been cited as potentially effective for addressing HCW wellbeing in the context of COVID-19 [21]. Generally, positive relationships with supervisors and coworkers have been associated with lower workplace stress [22]. Social support is recognized as a leading driver for this positive association [23]. Specifically, in Karasek's 1982 job strain model, coworker and supervisor support is understood to moderate associations between general occupational stressors, job control, and mental strain [24]. In a further examination of the role of supervisor and coworker support, after controlling for negative affect, high levels of supervisor support reduced the negative effects of job strain on level of job satisfaction, while coworker support mitigated the association between job strain and work performance [25]. Peer support models aid in the deconstruction of mental health stigma through increasing awareness and verbal expression of stress individually and as a team [23]. This is particularly relevant for HCWs whose occupation requires attention to and caring for others, often at their own expense [13]. Studies indicated that HCWs grappling with stressors from COVID-19 are interested in obtaining support, but few employees currently utilize emotional support resources [24].

The expansion of peer support services to address HCW needs during the COVID-19 pandemic represented an essential shift in occupational health [25–27]. In response to the pandemic, several hospitals and health systems have implemented hotlines and brief individual- or group-based supportive telehealth sessions for employees [28,29]. Support services have focused on the screening for and provision of mental health services at the individual, small group, and department levels, providing in-house stress-inoculation and resilience-promoting training, and, when clinically indicated, connecting HCWs to external referrals for ongoing care [25–27]. Additional services have included the dissemination of resources through online platforms (i.e., websites, apps), wherein employees can access online resources to support mental and physical health, psychosocial stressors (e.g., parenting and caregiving during COVID-19), and resilience maintenance. Initial qualitative findings from a hospital-based peer support program found shifts in organizational culture, staff skill building for recognizing and supporting coworkers in distress, and support for individuals already providing psychosocial peer support [30]. While the introduction and utilization of these services is a crucial step, the need for comprehensive, evidence-informed interventions remains.

A growing understanding of the negative impacts of COVID-19 on HCWs highlights the need for the development and evaluation of comprehensive, effective, and sustainable interventions to address the acute and long-term health needs of healthcare providers. To date, research exploring the use of empirically supported, individual-level treatments (e.g., cognitive behavioral therapy (CBT)) to improve mental health and wellbeing during the pandemic stop short of addressing resilience processes and contextual factors that impact wellbeing, such as perceptions of workplace support [17,26,28,29,31]. A systematic review published in the Cochrane Database determined there is a paucity of evidenced-based effectiveness research on interventions aimed at supporting the resilience and mental health of HCWs during and after disasters, asserting a strong need for studies examining comprehensive interventions for HCWs during COVID-19 [32]. The Systems Model theoretical framework delineates the various workplace levels that are impacted by stress and burnout, such as unit/team, leadership, and the healthcare industry [33]. For instance, at the workplace level, the Systems Model posits that stress and burnout will also be associated with negative consequences such as increased absenteeism, turnover, and lower engagement. The Systems Model also highlights individual and work system factors that might mediate stress and burnout, including coping strategies and coping self-efficacy, resilience, and organizational social support.

1.2. Stress First Aid

SFA is a peer support and self-care model involving the promotion of social support and resilience-building to mitigate stress reactions, individually and within teams [34–37]. Initially developed in the context of high-risk occupations (i.e., military service, fire and rescue, law enforcement), SFA has shown positive preliminary results for improving perceived ability to respond to behavioral health issues among teams [38]. SFA has, more recently, been adapted to meet the needs of HCWs. Specifically, SFA for HCWs is designed to promote peer-support interventions that have the potential to impact SFA self-efficacy, or the ability to recognize stress reactions in oneself and colleagues, encourage increased awareness and utilization of occupational wellbeing resources, improve perceptions of workplace support, and shift workplace stress and burnout levels. Stress First Aid is inherently compatible with Karasek's model; when demand is high, as was seen among HCWs during the pandemic, SFA seeks to increase control and social support within the workplace by enabling more effective communication around stressors and procurement of support, with the intent to decrease the negative impacts of job strain, including burnout. Stress First Aid acts as an effective mechanism for aligning efforts between programs as well with the goal of facilitating knowledge and awareness of employee supports such as EAP. This connection and cooperation promote resource-sharing and aid in the implementation of informed and efficient efforts in supporting our employees and the community.

SFA utilizes empirically informed elements that guide efforts for offering peer support actions in response to ongoing adversity, trauma, or disasters [34,39–41]. The five empirically informed elements include the promotion of a sense of (1) safety, (2) calm, (3) connectedness, (4) self- and community-efficacy, and (5) hope [42]. These directly map onto five of the seven Core Actions (7 C's) of SFA: cover, calm, connect, competence, and confidence. Check and coordinate are added as continuous core actions for situational monitoring of stress and linkage into care as needed. SFA's seven core actions directly impact coping self-efficacy, resilience, and perceptions of organizational support. Self-efficacy in terms of one's ability to cope with stress, has been linked with psychological stress itself as well as stress-related burnout, mental health symptoms, and workplace turnover intentions [43–46]. SFA is designed to generate increased individual and team coping and self-efficacy skills, in addition to improvements in leadership, workplace resilience, improved knowledge of and use of resources, and organizational social support. The SFA framework posits that stress injuries often result in decreased self-awareness of stress. This gap widens further for HCWs, who, by nature of the occupation, employ 'other' rather than 'self' focus and, therefore, are more likely to provide peer support than self-care. Thus, the provision of services within SFA includes direct peer support, wherein team members learn how to identify, communicate, and respond to stress. Initial research findings support the feasibility and acceptability of SFA among firefighter and nursing populations [40]. Figure 1 provides an overview of the anticipated impact of SFA.

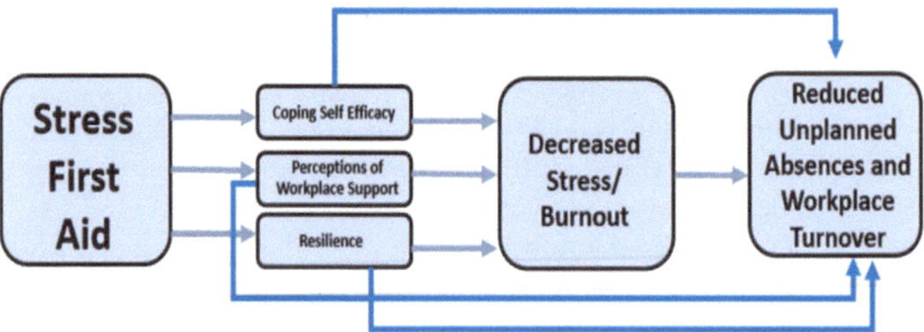

Figure 1. Impact of Stress First Aid.

The ability to test the impacts of SFA on the aforementioned constructs in the healthcare setting is important as there is limited research on its impact in the healthcare setting. SFA was developed with input from HCWs so that it would be feasible and acceptable for use with this population. SFA's flexible structure easily allows for these types of adaptations. For example, the lives of HCWs, particularly during a pandemic, are hectic and intensely busy. It was imperative that learning the basic framework of SFA would be a quick process. Healthcare settings can implement Stress First Aid setting-wide with 15 min briefings to raise awareness and educate about the importance of using SFA principles for stress mitigation. SFA champions or coaches can then train, mentor, and provide ongoing resources in their units. They also regularly assess the unit's overall stress levels on the stress continuum and shape the dialogue about the resources that exist to build resilience and capacity. This type of implementation is consistent with the EU Field Guide to Managing Complexity (and Chaos) in times of crisis, which recommends using a flexible approach built upon research- or theory-informed frameworks that empower adaptive actions in the context of complex public health situations [47]. It is also consistent with qualitative findings from a Cochrane review regarding work-related resilience interventions in the context of disease outbreaks [34,47]. This review noted that the successful implementation of programs depends upon flexible interventions that are culturally appropriate and adaptable to local needs. Also important are effective communication, cohesion through networks, a positive learning climate where team members feel valued and a part of the change process, and sufficient time and space for reflective thinking and evaluation. When HCWs are overwhelmed, they appreciate a "map" that can guide them through a variety of circumstances at whatever level they have the capacity for. SFA also allows for a long-term approach that includes multiple strategies for self-care strategies based on stress levels and current concerns, as well as sharing and learning from colleagues, consulting with mentors, and making time for the small actions that provide support.

1.3. Present Study

This study aims to address a gap in the current literature regarding the implementation and efficacy of a comprehensive, evidence-based peer support program in a cohort of HCWs at a large psychiatric hospital during the COVID-19 pandemic. The current study uses a longitudinal design to best capture fluctuations in outcomes over time and to understand both the short and longer-term potential impacts of SFA. The authors hypothesize that, over 12 months, SFA will be highly utilized, well-received, and efficacious in increasing participants' (a) self-efficacy, (b) resilience, (c) awareness and utilization of wellbeing resources, and (d) perceptions of organizational support. Secondary/distal outcomes will also be explored, including SFA's impact on stress and burnout.

2. Materials and Methods

2.1. Participants

To evaluate the effectiveness, implementation, and maintenance of the SFA program, 562 nurses and nursing support staff in a psychiatric hospital in the metropolitan New York area were given measures that were collected at baseline (before implementation), three, six, nine, and twelve months post SFA implementation to examine how each of the proximal and distal outcomes changes over the five-time points. Convenience sampling approaches were utilized to send each of the identified 562 HCWs the electronic survey. All data were collected and stored in a Research Electronic Data Capture (REDCap) database, which is a HIPAA-compliant database that requires a password and only allows access to specific individuals added by the CTSRR team. Data were collected between May 2021 and June 2022 via a REDCap electronic health questionnaire that was directly emailed to eligible participants. All study measurements, including SFA self-efficacy, wellbeing resource awareness/utilization, perceptions of organizational support, resilience, stress, and burnout, were contained within this questionnaire. Nurses and nursing staff who were working in an inpatient capacity at the hospital between May 2021 and June 2022 were included. Of the 562 nurses and nursing staff that were sent the survey, 150 participants completed the survey at baseline, 128 at the 3-month follow-up, 111 at the 6-month follow-up, 92 at the 9-month follow-up, and 116 at the 12-month follow-up. Table 1 shows the role breakdown of the 266 participants who completed the survey at any time point. Of the total sample, 77% consisted of mental health workers and registered nurses, with survey participants across 14 units of a single psychiatric hospital.

Table 1. Occupational role breakdown of unique respondents.

	Frequency	**Percent**
RN—Registered Nurse	96	36.1
Mental Health Nursing Staff	90	33.8
Unit Receptionist	20	7.5
ANM—Assistant Nurse Manager	14	5.3
Other	12	4.5
Nursing Administrator	9	3.4
NM—Nurse Manager	9	3.4
Scheduler	4	1.5
PCA—Patient Care Associate	3	1.1
Director	3	1.1
Nursing Attendant	2	0.8
Nurse Educator	2	0.8
NA—Nursing Assistant	1	0.4
Certified Nursing Assistant	1	0.4
Total	266	100

2.2. SFA Procedure

Implementation of SFA into a large hospital system entailed a four-step process that included preparation of teams, training, integration, and sustainment. SFA program staff included designated SFA Coaches expertly trained in SFA by two developers of SFA and project management support. The SFA program staff prepared a particular hospital team by orienting them to a process for implementation. Hospitals would identify an SFA site coordination and other personnel to support operations and report to executive sponsors. The SFA program team would help the hospital coordination team identify personnel that could facilitate training and team members that would support integration. Additional awareness raising activities such as briefings, newsletter announcements, and kickoff events were conducted during the preparation phase. SFA Coaches also prepared two levels of training- one for leaders and one for remaining staff. The coaches then provided master training for site trainers to deliver these trainings in more intimate settings to their staff and the coaches met regularly with the trainers to monitor them and provide feedback. SFA Coaches were also available to support responses to critical incidents. Critical to the SFA roll out was an integration process to embed SFA into daily team operations to moderate ongoing and exceptional occupational stressors across hospital units. Once about 50% of a site was trained, a 12-week integration phase was initiated in which leaders were asked to begin practices of "Raising Awareness", "Growing the Green", and "Stopping the Burn". This corresponded onto actions of posting signage, actively engaging in SFA practices (like a color check in and calming activities), and responding to elevated stress levels, respectively. The process of integration was monitored with self-report surveys through REDCap and a process of auditing. Lastly, sustainment was achieved by embedding check-ins on progress with SFA into hospital-wide administrative meetings, embedding practices like color checks into daily safety checks and weekly rounding, and by embedding training into new hire orientation so that new staff were trained. Progress with each step of the process was monitored with various process metrics.

2.3. Measures

We utilized a secure, HIPAA-compliant database, REDCap, for all data collection. The Human Resources department provided the contact information of all eligible nurses and nursing staff to the CTSRR team, which was then used to create a unique ID for each participant within REDCap. All study measurements were contained within the electronic baseline questionnaire, which was emailed directly to eligible participants through a secure and unique REDCap link at each time point. Responses are linked to a participant's record, where they can be tracked and analyzed.

2.3.1. Proximal Outcomes

Proximal outcomes include resilience, SFA self-efficacy, and perceptions of workplace support at each time point. The Connor–Davidson Resilience Scale 2© (CD-RISC 2) is a two-item scale that is useful as a brief measure of resilience to assess one's ability to adapt to change and bounce back after hardship [48]. The two items include "I am able to adapt when changes occur" and "I tend to bounce back after illness, injury, or other hardships". Each item on the CD-RISC 2 has responses that range between 1–5 (*1 = not true at all, 2 = rarely true, 3 = sometimes true, 4 = often true, 5 = true nearly all the time*). Item scores are summed for a total score range of 2–10 with higher scores on this screener indicative of greater levels of resilience.

The SFA self-efficacy score was developed by the research team to better understand SFA implementation (confidence in addressing the seven core actions of SFA), including confidence in one's ability to identify stress in co-workers, take action to reduce stress in co-workers, and link stressed co-workers into care. Items include "How confident are you in your ability to identify the level of risk/distress of an impacted person (i.e., self or co-worker) during and following a stressful event?"; "How confident are you in your knowledge of resources to link a stress-impacted person to needed supports?"; and

"How confident are you in your ability to take action to reduce the stress of an affected person/s (i.e., calm, cover, connect, competence, confidence)?". Item responses ranged from 1–4 (*1 = not confident, 2 = a little confident, 3 = confident, 4 = very confident*). Scores were summed for a total range of 3–12 with higher scores hypothesized to indicate greater SFA self-efficacy.

Perceived organizational support was measured using the Deployment Risk and Resilience Inventory-2 (DRRI-2), unit support subscale, adapted for use with frontline HCWs, which is a 12-item instrument with each item response between 1 and 5 (1 = strongly disagree; 5 = strongly agree) [45]. Total DRRI-2, Unit Support subscale score is created by summing each item rating with higher scores indicative of greater perceived social support from co-workers/unit members/unit leaders (range: 12–60). We aimed to measure co-worker and manager interest in employee wellbeing. Participants were asked whether "co-workers on my unit are interested in my wellbeing" and "the leaders of my unit are interested in my personal welfare", with responses between 1 and 5 (*1 = strongly disagree, 2 = disagree, 3 = neither agree nor disagree, 4 = agree, 5 = strongly agree*).

2.3.2. Distal Outcomes

Distal outcomes included stress and burnout levels. The Stress Continuum Model is a foundational part of the SFA model that helps with the assessment of stress response levels, ranging from *Green* (exceptionally low stress and stress reactions) through *Yellow* and *Orange* to *Red* (significant exposure to stress and experience of stress reactions) categories. Participants were asked to rate their level of stress for the past two weeks on a scale of 1–4, with *1* being the least stressful and *4* being the most stressful, and with the following corresponding colors: 1—green; 2—yellow; 3—orange; 4—red.

The Adapted Maslach Burnout Inventory (MBI) measured nine emotional exhaustion (EE) items, for example, "I feel burned out from my work", and five depersonalization (DP) items, including "I've noticed that I've become more callous (i.e., less sympathetic/compassionate) toward patients/customers". Burnout domains were evaluated on a scale of 1–5 (*1 = strongly disagree, 2 = disagree, 3 = neither agree nor disagree, 4 = agree, 5 = strongly agree*). For both subscales, higher mean scores are indicative of higher degrees of burnout.

To better understand the utilization of wellbeing resources, participants were asked to endorse a list of fourteen hospital-provided wellbeing resources that they are aware of and have utilized or directed a coworker to in the last three months. Examples of responses include the *Employee & Family Assistance Program (EAP), Team Lavender, Chaplain on-site support*, and the *Emotional Support Resource Call Center*.

Lastly, demographic data were collected including, position/job role, primary worksite/hospital, and primary work unit/department.

2.4. Statistical Analyses

All longitudinal analyses were conducted using the Generalized Estimating Equations (GEE) approach [49] under the assumption that the missing values were missing completely at random (MCAR). The working correlation structures considered for fitting GEE models for each variable were 'Independent', 'Exchangeable', 'AR-1', '3-Dependent', and 'Unstructured'. The criteria used for selecting the best covariance structure for each variable included both the Quasi Likelihood Information Criterion (QIC) values as well as a comparison between the model-based estimates and the empirical estimates of the correlation matrix [50]. Empirical estimators were obtained using the sandwich estimator [50]. PROC GENMOD in SAS version 9.4 was used for all GEE analyses. Marginal means for the continuous dependent variables and the odds for the binary dependent variables across all time points were estimated using the 'least squares means' option within the GENMOD procedure.

3. Results

3.1. Longitudinal Descriptive of Study Variables

Table 2 provides descriptive characteristics for participants' levels of self-efficacy, resilience, perceptions of organizational support, stress, burnout, and awareness of resources from baseline through 12-month follow-up.

Table 2. Outcome measures based on the least square means across all time Points.

	Baseline		3-Month Follow-Up		6-Month Follow-Up		9-Month Follow-Up		12-Month Follow-Up	
	M	**SE**	**M**	**SE**	**M**	**SE**	**M**	**SE**	**M**	**SE**
SFA Self-Efficacy *	9.10	0.15	9.15	0.13	9.36	0.14	9.21	0.14	9.54	0.14
Resilience **	8.28	0.11	8.13	0.12	8.13	0.13	8.17	0.12	8.47	0.11
Perceptions of Organizational Support	7.89	0.16	7.94	0.15	7.99	0.15	8.16	0.16	8.29	0.14
Stress	2.18	0.07	2.11	0.06	2.07	0.07	2.14	0.07	2.10	0.07
Burnout	5.11	0.13	4.99	0.14	5.20	0.17	5.85	0.15	4.82	0.14
Resource Awareness **	4.32	0.26	4.61	0.25	4.98	0.28	4.99	0.28	5.28	0.29
Any Resource Utilization (OR; 95% CI)	0.27	(0.18–0.40)	0.28	(0.19–0.43)	0.31	(0.20–0.48)	0.33	(0.21–0.53)	0.40	(0.27–0.60)

Note: * $p = 0.05$, ** $p < 0.01$.

3.2. Generalized Estimating Equation (GEE) Analysis of Study Variables

The overall comparisons of least square means were conducted across time points for each variable, presented in Figures 1–3. Proximal outcomes of and SFA self-efficacy ((χ^2, 4) = 9.45, $p = 0.051$) and resilience ((χ^2, 4) = 11.47, $p < 0.05$) significantly differed across time points. In addition, the average number of resources participants are aware of significantly differed across timepoints ((χ^2, 4) = 9.55, $p < 0.05$). All outcomes generally increased over time (see Figures 2–4).

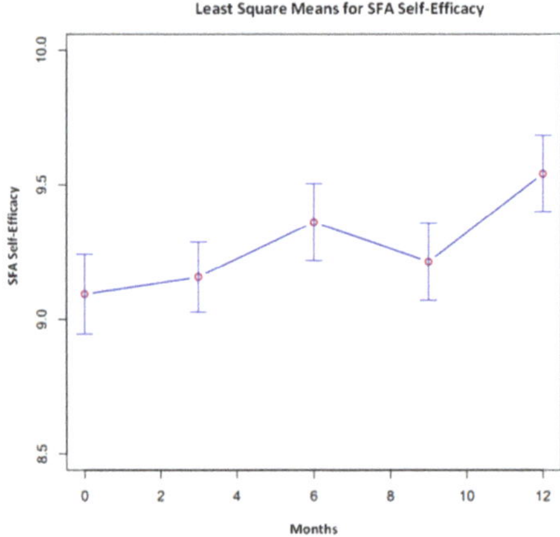

Figure 2. SFA self-efficacy across all time points.

Mean differences across time points did not significantly differ for the perception of organizational support ((χ^2, 4) = 5.67, $p = 0.225$), stress ((χ^2, 4) = 2.44, $p = 0.656$), burnout ((χ^2, 4) = 6.03, $p = 0.197$), or utilization of resources ((χ^2, 4) = 2.57, $p = 0.632$). When examined for change between the two-time points of baseline and 12 months, there are two additional significant results. Perceptions of organizational support significantly increased from baseline to 12 months (Mean difference = 0.396; $p = 0.0278$), and there was a borderline significant decrease in burnout (Mean difference = −0.287; $p = 0.057$).

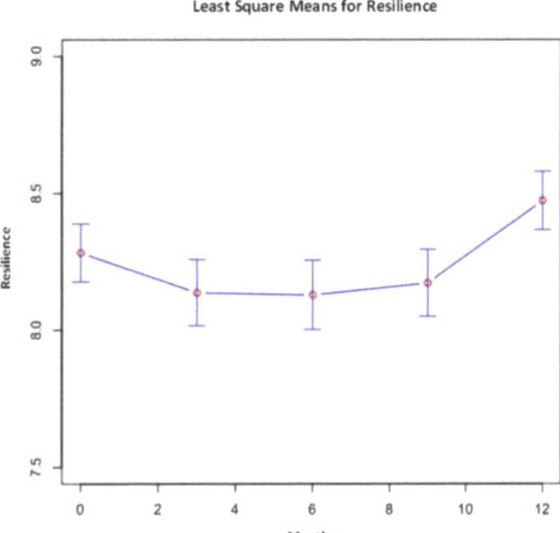

Figure 3. Resilience across all time points.

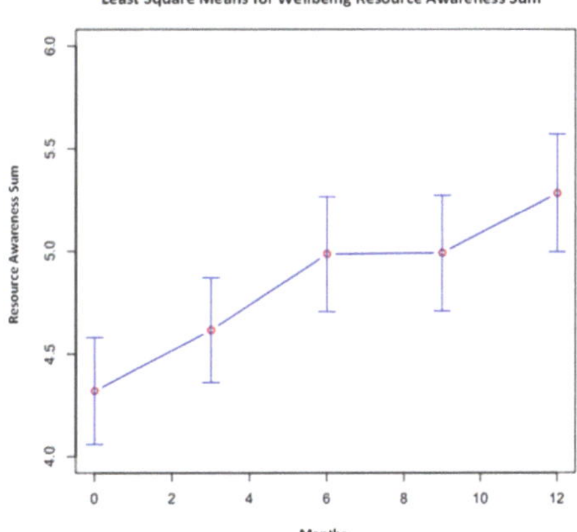

Figure 4. Awareness of wellbeing resources across all time points.

4. Discussion

Workplace stress, emphasized during the COVID-19 pandemic, remains a prominent issue among healthcare professionals. The authors of the present study sought to examine the use of Stress First Aid (SFA) as a targeted, self-care, and peer-support intervention to address healthcare worker wellbeing. Proximal outcomes were hypothesized to increase from baseline to 12-month follow-up, including self-efficacy, resource awareness and utilization, resilience, and perceptions of organizational support. At the same time, potential decreases in distal outcomes of stress and burnout were also explored. Generalized estimating equations (GEE) highlighted significant increases across three-month intervals from baseline to 12-month follow-up for self-efficacy, resource awareness, and resilience rates. Additionally, when considering only changes from baseline to 12-month follow-up, a significant increase was found in perceptions of organizational support and a borderline significant decrease in rates of burnout. In the present study, significance was not found for the utilization of resources or rates of stress.

This longitudinal study reveals unique shifts in the data across the 12 months. Rates of SFA self-efficacy increased, then decreased at month nine, followed by an increase at the twelve-month follow-up. A similar pattern emerged for rates of burnout, where, though not significant, participants reported higher rates at month nine before dropping in month twelve. Survey participation was also notably lower at month nine; however, response rates increased again at the twelve-month follow-up. There was a significant spike in COVID-19 cases in January 2022 [51], which may be reflected in the relative decrease in participation and feelings of self-efficacy at month nine, according to data that were collected in February and March 2022. Natural fluctuations in external stressors may account for some of the observed variability; however, a time-matched control group would be needed to further assess this hypothesis.

Another a point of interest is that the rates of resilience stayed fairly consistent during months 3–9, with a marked increase at the 12-month follow-up. Given the extended period of high-acuity work at the height of the pandemic for this sample of nurses, it is feasible that there was less capacity for resilience building initially, when the focus was required to be on addressing basic needs (i.e., fear of contamination) [52]. Resilience, as a construct, is generally measured as a state characteristic, not a trait characteristic, and may require more time under intervention exposure to see a quantifiable shift [53]. A larger sample size and dose/response analysis may be beneficial in determining this threshold for change. It is also likely that the threshold is variable-dependent, with SFA self-efficacy showing its greatest increase at the six-month time point.

4.1. Theoretical Implications

SFA aims to build protective factors to increase resilience and mitigate stress reactions. These factors can hopefully contribute longitudinally to decreased distress and burnout, and in this study, some significant change was found between baseline and 12-month follow-up for distal outcomes. However, the authors generally expected to see less change in measures of stress and burnout in the context of implementing SFA during a pandemic. Because significant change was found for distal outcomes even in this context, we are hopeful that in less stressful circumstances, the use of SFA can further contribute to a reduction in stress and burnout, mainly when the organization uses the model to build leadership's capacity to assess and address systemic issues which may have a more direct impact on stress and burnout. This focus should have an additive effect on the SFA goal of increasing awareness of stress injuries, helpful self-care, and peer support actions, resources, and access to care, as well as its goal of helping to identify, assist, and link into care those individuals who may be at risk for developing stress injuries that could predispose them to longer-term mental health conditions. Future studies should inspect the longer-term impact of SFA utilization across a healthcare system and potential company-wide cultural shifts in the discussion of stress and burnout.

4.2. Practical Implications

The findings from this study are significant because, in contrast to prior studies [33,35], this study protocol committed to implementing SFA in the recommended flexible way, with local SFA Coaches supported by more experienced supervisors. This approach yielded greater self-efficacy, resilience, and resource knowledge than studies with a more controlled, rigid format with less supervision [33,35]. This makes the current findings and guidelines more generalizable to the recommended implementation of SFA in high-stress work environments. Organizations that have the time and dedicated resources to follow up with multiple training opportunities, support from SFA Coaches, and multi-tiered ways to implement SFA actions should find better results in reduction in stress, increased coping self-efficacy, and perceived social and organizational support, mainly when initiated before public health emergencies rather than during a pandemic.

As noted previously, SFA is in alignment with implementation principles recommended for complex environments and public health emergencies, such as using research- or theory-informed frameworks to empower adaptive actions within a flexible approach to solve problems that a system cannot yet anticipate [47]. Building upon existing informal networks of support with a built-in level of trust allows for faster implementation, and greater trust allows HCWs to adapt actions to their moment-to-moment capacity. SFA, implemented this way, allows individuals and systems to define and move towards reduced stigma and enhanced communication about the ongoing necessary actions toward greater wellbeing and system effectiveness.

4.3. Limitations and Future Directions

The results of this study should be interpreted considering several limitations, with implications for future research. First, the present study had a relatively small sample size. Power was sufficient for the proposed statistical analyses; increasing the sample size in future studies may allow for group differentiation such as by role type, gender, and other variables. Additionally, a larger sample size or oversampling for additional demographic variables or variables of otherwise interest would have allowed for a more nuanced understanding of the effects of SFA among healthcare workers. The current study focused on healthcare workers in a behavioral health setting without specific control or comparison groups. For example, comparing study findings to control groups of other healthcare workers outside of a psychiatric setting may have been advantageous. While our outcomes assessment followed the proximal and distal outcomes of the Stress First Aid intervention, future repeated assessment of variables would strengthen causal inference and potential temporal mediation models. Whereas the present study focused on several proximal outcomes associated with occupational stress and burnout, future research might investigate additional outcomes. Research has found associations between occupational stress and other behavioral health outcomes, including substance use disorders and depression [54] among healthcare workers.

5. Conclusions

This study sought to extend the use of the peer support intervention, Stress First Aid, to healthcare workers during the COVID-19 pandemic. Findings provide preliminary support for the potential efficacy of SFA as a means of shifting culture around the discussion of emotional wellbeing, as well as an awareness and utilization of employee health resources. It will be important to continue to observe the potential impact of SFA during non-pandemic times as it has been used as a resilience-building, prevention strategy in military and first responder settings. It is possible that in less critical times, the impact on the more distal outcomes would be easier to observe because there would be fewer competing stressors. Further, results support the need to investigate implementation, dissemination and sustainability indicators as those will be integral in wider adoption of the model.

Author Contributions: Conceptualization, M.H.B., D.B., P.D., M.B.M., M.S., L.T., A.W.-M., P.W., R.J.W. and R.M.S.; methodology, H.M.C., P.S., G.Z. and R.M.S.; software, G.Z. and R.M.S.; validation, R.M.S.; data curation, G.Z. and R.M.S.; writing—original draft preparation, M.H.B., H.M.C., P.S. and R.M.S.; writing—review and editing, M.H.B., H.M.C., P.S., D.B., P.D., A.G., M.B.M., M.S., L.T., A.W.-M., G.Z., P.W., R.J.W. and R.M.S.; visualization, H.M.C. and G.Z.; supervision, M.H.B., M.S., L.T., P.W., R.J.W. and R.M.S.; project administration, M.H.B. and A.G. All authors have read and agreed to the published version of the manuscript.

Funding: This research received no external funding.

Institutional Review Board Statement: Ethical review and approval were waived for this study based on the determination that it constitutes non-human subjects research, specifically focusing on quality improvement-oriented programmatic evaluation.

Informed Consent Statement: Informed consent was waived as the study primarily focused on the analysis of existing program evaluation data and was determined to be non-human subjects' research.

Data Availability Statement: The data presented in this study are available by request and sent to the corresponding author.

Acknowledgments: We would like to express our gratitude to the healthcare providers for their unwavering commitment to their work. We thank them for their partnership in the Stress First Aid program.

Conflicts of Interest: The authors declare no conflicts of interest.

References

1. Ripp, J.; Peccoralo, L.; Charney, D. Attending to the Emotional Well-Being of the Health Care Workforce in a New York City Health System During the COVID-19 Pandemic. *Acad. Med.* **2020**, *95*, 1136–1139. [CrossRef]
2. Neto, M.L.R.; Almeida, H.G.; Esmeraldo, J.D.A.; Nobre, C.B.; Pinheiro, W.R.; de Oliveira, C.R.T.; da Costa Sousa, I.; Lima, O.M.M.L.; Lima, N.N.R.; Moreira, M.M.; et al. When Health Professionals Look Death in the Eye: The Mental Health of Professionals Who Deal Daily with the 2019 Coronavirus Outbreak. *Psychiatry Res.* **2020**, *288*, 112972. [CrossRef]
3. Kushal, A.; Kumar, S.; Mehta, M.; Singh, M. Study of Stress among Health Care Professionals: A Systemic Review. *Int. J. Res. Found. Hosp. Health Care Adm.* **2018**, *6*, 6–11. [CrossRef]
4. Choi, Y.-J. Psychological First-Aid Experiences of Disaster Health Care Workers: A Qualitative Analysis. *Disaster Med. Public Health Prep.* **2019**, *14*, 433–436. [CrossRef]
5. Arpacioglu, S.; Gurler, M.; Cakiroglu, S. Secondary Traumatization Outcomes and Associated Factors among the Health Care Workers Exposed to the COVID-19. *Int. J. Soc. Psychiatry* **2020**, *67*, 84–89. [CrossRef]
6. Carmassi, C.; Foghi, C.; Dell'Oste, V.; Cordone, A.; Bertelloni, C.A.; Bui, E.; Dell'Osso, L. PTSD Symptoms in Healthcare Workers Facing the Three Coronavirus Outbreaks: What Can We Expect after the COVID-19 Pandemic. *Psychiatry Res.* **2020**, *292*, 113312. [CrossRef] [PubMed]
7. Olfson, M.; Cosgrove, C.M.; Wall, M.M.; Blanco, C. Suicide Risks of Health Care Workers in the US. *JAMA* **2023**, *330*, 1161–1166. [CrossRef] [PubMed]
8. Shanafelt, T.; Ripp, J.; Trockel, M. Understanding and Addressing Sources of Anxiety among Health Care Professionals during the COVID-19 Pandemic. *JAMA* **2020**, *323*, 2133–2134. [CrossRef]
9. Firew, T.; Sano, E.D.; Lee, J.W.; Flores, S.; Lang, K.; Salman, K.; Greene, M.C.; Chang, B.P. Protecting the Front Line: A Cross-Sectional Survey Analysis of the Occupational Factors Contributing to Healthcare Workers' Infection and Psychological Distress during the COVID-19 Pandemic in the USA. *BMJ Open* **2020**, *10*, e042752. [CrossRef]
10. Fateminia, A.; Hasanvand, S.; Goudarzi, F.; Mohammadi, R. Post-Traumatic Stress Disorder among Frontline Nurses during the COVID-19 Pandemic and Its Relationship with Occupational Burnout. *Iran. J. Psychiatry* **2022**, *17*, 436–445. [CrossRef] [PubMed]
11. U.S. Census Bureau. Women Hold 76% of All Health Care Jobs, Gaining in Higher-Paying Occupations. Census.gov. Available online: https://www.census.gov/library/stories/2019/08/your-health-care-in-womens-hands.html (accessed on 21 February 2023).
12. Labrague, L.J.; de Los Santos, J.A.A. Fear of COVID-19, Psychological Distress, Work Satisfaction and Turnover Intention among Frontline Nurses. *J. Nurs. Manag.* **2020**, *29*, 395–403. [CrossRef] [PubMed]
13. Labrague, L.J.; De Los Santos, J.A.A. COVID-19 Anxiety among Front-Line Nurses: Predictive Role of Organisational Support, Personal Resilience and Social Support. *J. Nurs. Manag.* **2020**, *28*, 1653–1661. [CrossRef] [PubMed]
14. Said, R.M.; El-Shafei, D.A. Occupational Stress, Job Satisfaction, and Intent to Leave: Nurses Working on Front Lines during COVID-19 Pandemic in Zagazig City, Egypt. *Environ. Sci. Pollut. Res.* **2020**, *28*, 8791–8801. [CrossRef] [PubMed]
15. Perego, G.; Cugnata, F.; Brombin, C.; Milano, F.; Mazzetti, M.; Taranto, P.; Preti, E.; Di Pierro, R.; De Panfilis, C.; Madeddu, F.; et al. Analysis of Healthcare Workers' Mental Health during the COVID-19 Pandemic: Evidence from a Three-Wave Longitudinal Study. *J. Health Psychol.* **2023**, *28*, 1279–1292. [CrossRef] [PubMed]
16. Vacca, A.; Minò, M.V.; Longo, R.; Lucisani, G.; Solomita, B.; Franza, F. The Emotional Impact on Mental Health Workers in the Care of Patients with Mental Disorders in the Pandemic and Post COVID-19 Pandemic: A Measure of "Burnout" and "Compassion Fatigue". *Psychiatr. Danub.* **2023**, *35*, 292–295. [PubMed]
17. Kang, L.; Ma, S.; Chen, M.; Yang, J.; Wang, Y.; Li, R.; Yao, L.; Bai, H.; Cai, Z.; Xiang Yang, B.; et al. Impact on Mental Health and Perceptions of Psychological Care among Medical and Nursing Staff in Wuhan during the 2019 Novel Coronavirus Disease Outbreak: A Cross-Sectional Study. *Brain Behav. Immun.* **2020**, *87*, 11–17. [CrossRef]
18. Mahaffey, B.L.; Mackin, D.M.; Rosen, J.; Schwartz, R.M.; Taioli, E.; Gonzalez, A. The Disaster Worker Resiliency Training Program: A Randomized Clinical Trial. *Int. Arch. Occup. Environ. Health* **2020**, *94*, 9–21. [CrossRef]
19. Schwartz, R.M.; McCann-Pineo, M.; Bellehsen, M.; Singh, V.; Malhotra, P.; Rasul, R.; Corley, S.S.; Jan, S.; Parashar, N.; George, S.; et al. The Impact of Physicians' COVID-19 Pandemic Occupational Experiences on Mental Health. *J. Occup. Environ. Med.* **2021**, *64*, 151–157. [CrossRef]
20. Singh, V.; Young, J.Q.; Malhotra, P.; McCann-Pineo, M.; Rasul, R.; Corley, S.S.; Yacht, A.C.; Friedman, K.; Barone, S.; Schwartz, R.M. Evaluating Burnout during the COVID-19 Pandemic among Physicians in a Large Health System in New York. *Arch. Environ. Occup. Health* **2022**, *77*, 819–827. [CrossRef]
21. Shechter, A.; Diaz, F.; Moise, N.; Anstey, D.E.; Ye, S.; Agarwal, S.; Birk, J.L.; Brodie, D.; Cannone, D.E.; Chang, B.; et al. Psychological Distress, Coping Behaviors, and Preferences for Support among New York Healthcare Workers during the COVID-19 Pandemic. *Gen. Hosp. Psychiatry* **2020**, *66*, 1–8. [CrossRef]
22. Quinn, A.; Ji, P.; Nackerud, L. Predictors of Secondary Traumatic Stress among Social Workers: Supervision, Income, and Caseload Size. *J. Soc. Work* **2018**, *19*, 504–528. [CrossRef]
23. Dennis, C.-L. Peer Support within a Health Care Context: A Concept Analysis. *Int. J. Nurs. Stud.* **2003**, *40*, 321–332. [CrossRef]
24. Brooks, C.D.; Ling, J. "Are We Doing Enough": An Examination of the Utilization of Employee Assistance Programs to Support the Mental Health Needs of Employees During the COVID-19 Pandemic. *J. Insur. Regul.* **2020**, *39*, 1–34. [CrossRef]

25. Mellins, C.A.; Mayer, L.E.S.; Glasofer, D.R.; Devlin, M.J.; Albano, A.M.; Nash, S.S.; Engle, E.; Cullen, C.; Ng, W.Y.K.; Allmann, A.E.; et al. Supporting the Well-Being of Health Care Providers during the COVID-19 Pandemic: The CopeColumbia Response. *Gen. Hosp. Psychiatry* **2020**, *67*, 62–69. [CrossRef]

26. Albott, C.S.; Wozniak, J.R.; McGlinch, B.P.; Wall, M.H.; Gold, B.S.; Vinogradov, S. Battle Buddies: Rapid Deployment of a Psychological Resilience Intervention for Health Care Workers During the Coronavirus Disease 2019 Pandemic. *Anesth. Analg.* **2020**, *131*, 43–54. [CrossRef] [PubMed]

27. DePierro, J.; Katz, C.L.; Marin, D.; Feder, A.; Bevilacqua, L.; Sharma, V.; Hurtado, A.; Ripp, J.; Lim, S.; Charney, D. Mount Sinai's Center for Stress, Resilience and Personal Growth as a Model for Responding to the Impact of COVID-19 on Health Care Workers. *Psychiatry Res.* **2020**, *293*, 113426. [CrossRef] [PubMed]

28. Weiner, L.; Berna, F.; Nourry, N.; Severac, F.; Vidailhet, P.; Mengin, A.C. Efficacy of an Online Cognitive Behavioral Therapy Program Developed for Healthcare Workers during the COVID-19 Pandemic: The REduction of STress (REST) Study Protocol for a Randomized Controlled Trial. *Trials* **2020**, *21*, 870. [CrossRef] [PubMed]

29. Rosenberg, A.R. Cultivating Deliberate Resilience during the Coronavirus Disease 2019 Pandemic. *JAMA Pediatr.* **2020**, *174*, 817–818. [CrossRef] [PubMed]

30. Anderson, G.S.; Di Nota, P.M.; Groll, D.; Carleton, R.N. Peer Support and Crisis-Focused Psychological Interventions Designed to Mitigate Post-Traumatic Stress Injuries among Public Safety and Frontline Healthcare Personnel: A Systematic Review. *Int. J. Environ. Res. Public Health* **2020**, *17*, 7645. [CrossRef] [PubMed]

31. Buselli, R.; Corsi, M.; Baldanzi, S.; Chiumiento, M.; Del Lupo, E.; Dell'Oste, V.; Bertelloni, C.A.; Massimetti, G.; Dell'Osso, L.; Cristaudo, A.; et al. Professional Quality of Life and Mental Health Outcomes among Health Care Workers Exposed to SARS-CoV-2 (COVID-19). *Int. J. Environ. Res. Public Health* **2020**, *17*, 6180. [CrossRef]

32. Blake, H.; Bermingham, F.; Johnson, G.; Tabner, A. Mitigating the Psychological Impact of COVID-19 on Healthcare Workers: A Digital Learning Package. *Int. J. Environ. Res. Public Health* **2020**, *17*, 2997. [CrossRef] [PubMed]

33. National Academies of Sciences, Engineering, and Medicine; National Academy of Medicine. Committee on Systems Approaches to Improve Patient Care by Supporting Clinician Well-Being. In *Taking Action Against Clinician Burnout: A Systems Approach to Professional Well-Being*; National Academies Press (US): Washington, DC, USA, 2019.

34. Pollock, A.; Campbell, P.; Cheyne, J.; Cowie, J.; Davis, B.; McCallum, J.; McGill, K.; Elders, A.; Hagen, S.; McClurg, D.; et al. Interventions to Support the Resilience and Mental Health of Frontline Health and Social Care Professionals during and after a Disease Outbreak, Epidemic or Pandemic: A Mixed Methods Systematic Review. *Cochrane Libr.* **2020**, *11*, CD013779. [CrossRef]

35. Nash, W.P.; Westphal, R.J.; Watson, P.J.; Litz, B.T. *Combat and Operational Stress First Aid: Caregiver Training Manual*; U.S. Navy, Bureau of Medicine and Surgery: Washington, DC, USA, 2010.

36. Watson, P.; Taylor, V.; Gist, R.; Elvander, E.; Leto, F.; Martin, B.; Tanner, J.; Vaught, D.; Nash, W.; Westphal, R.; et al. *Stress First Aid for Firefighters and Emergency Medical Services Personnel*; National Fallen Firefighters Foundation: Emmitsburg, MD, USA, 2015.

37. Watson, P.J.; Brymer, M.J.; Bonanno, G.A. Postdisaster Psychological Intervention since 9/11. *Am. Psychol.* **2011**, *66*, 482–494. [CrossRef] [PubMed]

38. Jahnke, S.A.; Watson, P.; Leto, F.; Jitnarin, N.; Kaipust, C.M.; Hollerbach, B.S.; Haddock, C.K.; Poston, W.S.C.; Gist, R. Evaluation of the Implementation of the NFFF Stress First Aid Intervention in Career Fire Departments: A Cluster Randomized Controlled Trial. *Int. J. Environ. Res. Public Health* **2023**, *20*, 7067. [CrossRef] [PubMed]

39. Dong, L.; Meredith, L.S.; Farmer, C.M.; Ahluwalia, S.C.; Chen, P.G.; Bouskill, K.; Han, B.; Qureshi, N.; Dalton, S.; Watson, P.; et al. Protecting the Mental and Physical Well-Being of Frontline Health Care Workers during COVID-19: Study Protocol of a Cluster Randomized Controlled Trial. *Contemp. Clin. Trials* **2022**, *117*, 106768. [CrossRef]

40. McLean, C.P.; Betsworth, D.; Bihday, C.; Daman, M.C.; Davis, C.A.; Kaysen, D.; Rosen, C.S.; Saxby, D.; Smith, A.E.; Spinelli, S.; et al. Helping the Helpers: Adaptation and Evaluation of Stress First Aid for Healthcare Workers in the Veterans Health Administration During the COVID-19 Pandemic. *Workplace Health Saf.* **2023**, *71*, 162–171. [CrossRef]

41. Watson, P.J. Application of the Five Elements Framework to the Covid Pandemic. *Psychiatry MMC* **2021**, *84*, 415–429. [CrossRef]

42. Hobfoll, S.E.; Watson, P.; Bell, C.C.; Bryant, R.A.; Brymer, M.J.; Friedman, M.J.; Friedman, M.; Gersons, B.P.R.; de Jong, J.T.V.M.; Layne, C.M.; et al. Five Essential Elements of Immediate and Mid-Term Mass Trauma Intervention: Empirical Evidence. *Psychiatry Interpers. Biol. Process.* **2007**, *70*, 283–315. [CrossRef]

43. Fida, R.; Laschinger, H.K.S.; Leiter, M.P. The Protective Role of Self-Efficacy against Workplace Incivility and Burnout in Nursing: A Time-Lagged Study. *Health Care Manag. Rev.* **2018**, *43*, 21–29. [CrossRef]

44. Laschinger, H.K.S.; Borgogni, L.; Consiglio, C.; Read, E. The Effects of Authentic Leadership, Six Areas of Worklife, and Occupational Coping Self-Efficacy on New Graduate Nurses' Burnout and Mental Health: A Cross-Sectional Study. *Int. J. Nurs. Stud.* **2015**, *52*, 1080–1089. [CrossRef]

45. Molero, M.D.M.; Pérez-Fuentes, M.D.C.; Gázquez, J.J. Analysis of the Mediating Role of Self-Efficacy and Self-Esteem on the Effect of Workload on Burnout's Influence on Nurses' Plans to Work Longer. *Front. Psychol.* **2018**, *9*, 2605. [CrossRef]

46. Pisanti, R.; van der Doef, M.; Maes, S.; Lombardo, C.; Lazzari, D.; Violani, C. Occupational Coping Self-Efficacy Explains Distress and Well-Being in Nurses beyond Psychosocial Job Characteristics. *Front. Psychol.* **2015**, *6*, 1143. [CrossRef]

47. Snowden, D.; Rancati, A. Managing Complexity (and Chaos) in Times of Crisis. A Field Guide for Decision Makers Inspired by the Cynefin Framework. *Publ. Off. Eur. Union* **2021**, 20–34. [CrossRef]

48. Connor, K.M.; Davidson, J.R.T. Development of a New Resilience Scale: The Connor-Davidson Resilience Scale (CD-RISC). *Depress. Anxiety* **2003**, *18*, 76–82. [CrossRef]
49. Liang, K.-Y.; Zeger, S.L. Longitudinal Data Analysis Using Generalized Linear Models. *Biometrika* **1986**, *73*, 13–22. [CrossRef]
50. Fitzmaurice, G.M.; Laird, N.M.; Ware, J.H. *Applied Longitudinal Analysis*, 2nd ed.; Wiley: Hoboken, NJ, USA, 2011.
51. Department of Health. Positive Tests over Time, by Region and County. Available online: https://coronavirus.health.ny.gov/positive-tests-over-time-region-and-county (accessed on 16 January 2024).
52. Ghanimi, H.M.A.; Yasear, S.A. A Cognitive Agent Model of Burnout for Front-Line Healthcare Professionals in Times of COVID-19 Pandemic. *Int. J. Intell. Eng. Syst.* **2022**, *15*, 348–360. [CrossRef]
53. Child, L.E.; Medvedev, O.N. Investigating State and Trait Aspects of Resilience Using Generalizability Theory. *Psychol* **2023**. [CrossRef]
54. Sriharan, A.; Ratnapalan, S.; Tricco, A.C.; Lupea, D.; Ayala, A.P.; Pang, H.; Lee, D.D. Occupational Stress, Burnout, and Depression in Women in Healthcare During COVID-19 Pandemic: Rapid Scoping Review. *Front. Glob. Women's Health* **2020**, *1*, 596690. [CrossRef]

International Journal of
Environmental Research and Public Health

MDPI

Article

Mentalizing, Resilience, and Mental Health Status among Healthcare Workers during the COVID-19 Pandemic: A Cross-Sectional Study

Teodora Safiye [1,*], Medo Gutić [1,2,3], Jakša Dubljanin [1,4], Tamara M. Stojanović [5], Draško Dubljanin [1,6], Andreja Kovačević [1,7], Milena Zlatanović [1,8], Denis H. Demirović [9], Nemanja Nenezić [8] and Ardea Milidrag [1]

[1] Faculty of Medical Sciences, University of Kragujevac, Svetozara Markovića 69, 34000 Kragujevac, Serbia
[2] Public Health Institution Health Center "Dr Branko Zogovic", Hridska bb, 84325 Plav, Montenegro
[3] Faculty of Medicine, University of Belgrade, Dr Subotića 8, 11000 Belgrade, Serbia
[4] Department of Cardiology, University Clinical Hospital Center Zemun, Vukova 9, 11080 Belgrade, Serbia
[5] Faculty of Philology and Arts, University of Kragujevac, Jovana Cvijića bb, 34000 Kragujevac, Serbia
[6] Department of Pulmonology, University Clinical Hospital Center Zvezdara, Dimitrija Tucovića 161, 11120 Belgrade, Serbia
[7] Department of Cardiopulmonary Rehabilitation, Institute for Rehabilitation Belgrade, Sokobanjska 17, 11000 Belgrade, Serbia
[8] Department of Medical Studies Ćuprija, Academy of Educational and Medical Vocational Studies Kruševac, Bulevar Vojske bb, 35230 Ćuprija, Serbia
[9] Department of Psychology, Faculty of Philoshopy and Arts, State University of Novi Pazar, Vuka Karadžića 9, 36300 Novi Pazar, Serbia
*   Correspondence: teodoras0306@gmail.com

Abstract: The COVID-19 pandemic has caused unprecedented stress on healthcare professionals worldwide. Since resilience and mentalizing capacity play very important preventive roles when it comes to mental health, the main goal of this study was to determine whether the capacity for mentalizing and resilience could explain the levels of depression, anxiety, and stress among healthcare workers during the COVID-19 pandemic. The study was conducted in Serbia on a sample of 406 healthcare workers (141 doctors and 265 nurses) aged 19 to 65 (M = 40.11, SD = 9.41). The participants' mental health status was evaluated using the Depression, Anxiety, and Stress Scale—DASS-42. The Reflective Functioning Questionnaire was used to evaluate the capacity for mentalizing. Resilience was assessed using the Brief Resilience Scale. The results of the correlation analysis showed that there were negative correlations between resilience and all three dimensions of mental health status: depression, anxiety, and stress. Hypermentalizing was negatively correlated with depression, anxiety, and stress, while hypomentalizing was positively correlated. Hierarchical linear regression analysis showed that both resilience and hypermentalizing were significant negative predictors of depression, anxiety, and stress, and that hypomentalizing was a significant positive predictor of depression, anxiety, and stress. Furthermore, socioeconomic status was a significant negative predictor of depression, anxiety, and stress. Marital status, number of children, and work environment were not statistically significant predictors of any of the three dimensions of mental health status among the healthcare workers in this study. There is an urgent need to establish and implement strategies to foster resilience and enhance the capacity for mentalizing among healthcare workers in order to minimize the devastating effects of the COVID-19 pandemic on mental health.

Keywords: COVID-19; mental health status; mentalizing; depression; anxiety; stress; resilience; Serbia; doctors; nurses

Citation: Safiye, T.; Gutić, M.; Dubljanin, J.; Stojanović, T.M.; Dubljanin, D.; Kovačević, A.; Zlatanović, M.; Demirović, D.H.; Nenezić, N.; Milidrag, A. Mentalizing, Resilience, and Mental Health Status among Healthcare Workers during the COVID-19 Pandemic: A Cross-Sectional Study. *Int. J. Environ. Res. Public Health* **2023**, *20*, 5594. https://doi.org/10.3390/ijerph20085594

Academic Editor: Kevin M. Kelly

Received: 3 February 2023
Revised: 1 April 2023
Accepted: 13 April 2023
Published: 20 April 2023

1. Introduction

Severe acute respiratory syndrome coronavirus 2, also referred to as SARS-CoV-2, is a highly contagious, morbific virus that belongs to the group of coronaviruses. With its outbreak in China in 2019, there was a pandemic of an acute respiratory disease called 'coronavirus disease 2019' (COVID-19), which resulted in endangering the health of the population, as well as public safety. The World Health Organization announced a pandemic in March 2020 due to the involvement of over 110 countries and territories around the world [1]. In addition to the fact that COVID-19 caused changes in daily lifestyle, it had devastating effects on both the physical and mental health of individuals [2,3]; a special impact was noted in the field of mental health of healthcare workers (HCWs) worldwide [4]. The consequences of the COVID-19 pandemic on mental health are most often manifested in the form of increased symptoms of stress, anxiety, frustration, and depression. Common psychological reactions associated with changing lifestyles during a pandemic are generalized fear and anxiety, which are usually caused by information about the easy transmissibility of the virus and rapid escalation of new cases of infection, as well as previous information from the media that cause restlessness and a sense of uncertainty [3]. The absence of social interaction due to isolation measures additionally contributes to the experience of stress since social activities and support help regulate emotions, overcome stress, and maintain resilience when facing difficult life events [2].

An increased level of distress is associated with working in healthcare, even under regular circumstances, and in this context, anxiety, depression, sleep disorders, and burnout syndrome among healthcare workers were described [5,6]. However, throughout the course of fighting against the pandemic, healthcare workers were at risk of facing enormous amounts of stress, as a large number of medical personnel became infected due to the outbreak of SARS-CoV-2, which further increased the psychological stress of their colleagues [1]. In addition, they faced not only a greater risk of infection and the fear of infection and the spread of the virus to their loved ones but also emotional disorders, sleep problems, isolation, lack of contact with their families, extended shifts, and physical exhaustion [4,7]. Healthcare workers are responsible for other people's lives; therefore, all these psychological burdens, especially among medical staff on the frontline, can lead to a strong negative impact on their mental health [8]. There is increasing evidence to suggest that COVID-19 may be an independent risk factor for stress in healthcare workers. In a review of six studies, Spoorthy et al. reported that healthcare workers experienced significant levels of stress, anxiety, depression, and insomnia due to the COVID-19 pandemic [9]. It was also shown that in addition to psychological variables, such as poor social support and low self-efficacy, certain sociodemographic variables, such as gender, profession, age, and place of work negatively affected mental health, more precisely, they were associated with increased symptoms of stress, anxiety, and depression among healthcare workers [9]. It is considered that resilience [6,10] and good mentalizing capacity [11] play very important roles in reducing these mental health symptoms and in preserving mental health while facing crisis situations, such as the pandemic.

When we talk about psychological resilience, we are talking about the ability of people to more easily tolerate increased levels of stress and to function adequately in difficult circumstances. In the literature, it is most often stated that resilience is a general personality trait that implies the capacity for effective adaptation and the capacity to sustain or regenerate mental health under challenging circumstances [12]. Smith et al. believed that resilience is not only a process of manifesting positive perspectives and coping mechanisms but also the individual's capacity to quickly and efficiently recover from stressful experiences [13]. Multidisciplinarity in the approach to this problem also led to the evolution and change of definitions known so far, along with the evolution of the scientific understanding and knowledge of this issue. Thus, over time, several sources of resilience that are interconnected were identified: (1) Individual characteristics, such as openness, extraversion, agreeableness, and cooperation; internal locus of control; optimism; cognitive flexibility; emotional regulation; hope; resourcefulness; and adaptability. (2) Biological factors—an unfavorable early environment can affect the size, neuronal networks, receptor sensitivity, and the synthesis and uptake of neurotransmitters in the developing brain. These structural alterations in the brain can significantly increase or decrease susceptibility to psychopathology in the future, as well as the capacity to control negative emotions and, consequently, resilience to adversity. These changes can lead to the impairment of a person's ability to cope with emotions. (3) Environmental and systemic factors—social support from family and peers, secure bond with the mother, a stable family, an appropriate relationship with a non-abusive parent, effective parenting techniques, etc. [14]. Well-developed resilience was cited even before the COVID-19 pandemic as a trait that could enable healthcare workers to recover more easily from various difficulties and could be acquired through an appropriate training program [6].

Recently, the capacity for mentalizing was cited as an important characteristic of resilience, which implies the human ability to understand one's own and others' intentional mental states, which affects the general resilience of the personality, i.e., the ability to successfully adapt to challenges and stress. This aspect of reflective personality functioning can also be characterized as a person's ability to understand himself/herself and others, both in terms of subjective mental states and in terms of mental processes. In essence, the concept of mentalizing refers to how an individual explains his behavior and the behavior of others through the interpretation of the reasons, causes, and motives of events [15]. When we talk about the process of mentalizing itself, we are talking about the interdependence between subjective understanding of the mental states of ourselves and others, which affects our behavior and allows us to feel in control of our thinking and the way we act, as well as the way we take in, analyze and interpret social information from the environment around us [16]. However, as stated by some authors [17], in the process of mentalizing, we can distinguish between two qualitatively different, wrong ways of interpreting signals from the environment. These are hypomentalizing and hypermentalizing. Making too many assumptions about intentional mental states, without thinking critically about their truth, is called hypermentalizing. It manifests as excessive confidence in the veracity of one's own views about the nature of the mental processes that underlie one's actions and arises as a result of the individual's incorrect beliefs that other persons have similar or the same intentional mental states. In contrast, hypomentalizing refers to the lack of assumptions about the mental states that determine behavior in interpersonal interaction, which manifests itself as uncertainty regarding the accurate assessment of the mental states underlying some behavior [17]. Bearing in mind the mentioned characteristics of mentalizing, it is crucial to implement preventive or early interventions that encourage the ability to mentalize, which contributes to increasing resilience to life stressors, and thus, protecting individuals from the influence of disturbing factors and preserving their mental health [18]. This is especially important nowadays when every individual and society is faced with numerous stressful situations, such as the COVID-19 pandemic and its consequences.

With all of this in mind, the primary objective of this study was to determine whether mentalizing capacity and resilience could account for the levels of stress, anxiety, and depression experienced by healthcare professionals during the COVID-19 epidemic in Serbia. We hypothesized that significant relationships would be found between resilience, as well as impairments in mentalizing capacity, and the mental health status of HCWs (operationalized through levels of depression, anxiety, and stress). As far as we know, this study is one of the few that examined the specificity of the protective impact of mentalizing capacity on mental health in a sample of healthcare workers facing the COVID-19 pandemic.

2. Materials and Methods

2.1. Study Design, Participants, and Procedures

A cross-sectional study design was adopted. The current study is part of a large, self-funded project entitled "Burnout syndrome and indicators of mental health of workers during the COVID-19 pandemic in Serbia", led by Teodora Safiye. The Institutional Review Board of the University of Belgrade's Faculty of Philosophy, Department of Psychology, approved the project (approval number: #2021-58). The protocols used in the present study adhered to the Declaration of Helsinki's guidelines for medical research with human individuals [19]. Prior to the study's recruitment, each subject provided a written statement of informed consent. Each participant was given comprehensive information on their rights to inquire about their involvement in the study and to withdraw at any time without consequence. The study report did not include the names of the participants, and all data gathered from them was protected so that just the investigators could access it. The Raosoft Sample Size Calculator (available online: http://www.raosoft.com/samplesize.html (accessed on 1 May 2021)) was used to determine the necessary sample size. A sample of 377 respondents was estimated using the following assumptions: a 5% margin of error, a 95% confidence interval, and a population size of 20,000. Doctors and nurses from Serbia who were living in Serbia, actively caring for patients at the University Clinical Center Kragujevac at the time of the study, and between the ages of 18 and 65 were eligible to participate in the study. After the approval of the Ethics Committee of the University Clinical Center Kragujevac, we started gathering data using the paper-and-pencil survey method, in which survey participants had to complete a paper-based survey by hand. The research's aims were disclosed to possible participants in Serbian from the very beginning of the application of the questionnaire, and participation was voluntary, free of charge, and with informed consent. At the same time, the respondents were guaranteed the confidentiality and anonymity of the data obtained. The research was carried out between 15 July 2021 and 5 February 2022 at the University Clinical Center Kragujevac in Serbia during the height of the COVID-19 outbreak in Serbia [20].

2.2. Measures

The Depression, Anxiety, and Stress Scale—DASS-42 [21] was used to evaluate the participants' mental health status, namely, to measure depression, anxiety, and stress. The DASS-42 scale consists of 42 items and includes three subscales: depression, anxiety, and stress. According to the authors, the items of the depression subscale refer to experiences of dysphoria (e.g., "I felt sad and miserable"), anhedonia (e.g., "I couldn't seem to experience any positive feelings at all"), hopelessness (e.g., "I felt that I had nothing to look forward to"), devaluing oneself (e.g., "I felt I wasn't worth much as a person"), and devaluing life (e.g., "I felt that life wasn't worthwhile"); the anxiety subscale refers to experiences of apprehension, high tension, and helplessness (e.g., "I felt I was close to panic"); and the stress subscale refers to high arousal of the organism and negative emotions arising as a result of unpleasant or threatening events (e.g., "I found it difficult to relax"). The score on the depression subscale is calculated by adding the points from questions 3, 5, 10, 13, 16, 17, 21, 24, 26, 31, 34, 37, 38, and 42. The score on the anxiety subscale is calculated by adding the points from questions 2, 4, 7, 9, 15, 19, 20, 23, 25, 28, 30, 36, 40, and 41. The score on the stress subscale is calculated by adding the points from questions 1, 6, 8, 11, 12, 14, 18, 22, 27, 29, 32, 33, 35, and 39. The total score on each of the three mentioned subscales can range from 0 to 42. Respondents used a 4-point Likert-type scale, from 0—not at all to 3—mostly or almost always, to assess their degree of agreement without dwelling too much on the items. Since this research examined depression, anxiety, and stress as indicators of the respondents' mental health status (not as personality traits), the respondents were instructed to carefully read each item and mark the answer that best described how they felt in the previous 7 days [21]. The original long form of the DASS-42 questionnaire has had very good reliability, where the Cronbach alpha coefficient was above 0.8 in previous studies [22,23].

Resilience was measured using the Brief Resilience Scale (BRS), which was developed by Smith et al. [13]. This scale assesses the construct of resilience, which is understood as an individual's ability to cope with obstacles from the environment and recover from stressful circumstances. The short resilience scale has very good reliability, where the Cronbach alpha coefficient was above 0.8 in previous research [6,13]. Six items make up the one-dimensional, brief resilience scale. Resilience is endorsed by three items (e.g., "I typically come through challenging times with little trouble"), while it is dismissed by three items with reverse scoring (e.g., "I have a hard time making it through stressful events"). The mean of all six items represents the final score on this scale. Respondents had the option of selecting one response on a five-point Likert-type scale, from 1—completely false to 5—completely true [13]. A shortened version of the Reflective Functioning Questionnaire (RFQ-8) [17], which is advised for research to prevent subject fatigue, was used to assess mentalizing capacity. The RFQ-8 has two subscales: (1) the subscale of certainty in one's own assessment of mental states (RFQ-c), where high scores denote the phenomenon of hypermentalizing, and (2) the subscale of uncertainty in one's own ability to assess one's own and others' mental states (RFQ-u), which denotes the phenomenon of hypomentalizing. The RFQ-c subscale measures how strongly a person disagrees with statements like "People's thoughts are a mystery to me" in order to determine how confident a person is in their ability to effectively evaluate both their own and others' mental states. The RFQ subscale assesses a person's level of insecurity regarding their capacity to judge both their own and other people's mental conditions. The degree to which a person agrees with phrases like "I don't always know why I do what I do" is used to measure it. Low values indicate optimal mentalizing, while high scores indicate hypomentalizing [24,25]. Responses are rated on a seven-point Likert scale, with 1 representing "I do not agree at all" and 7 representing "I completely agree". The responses received from respondents on both RFQ-8 subscales could range from 0 to 3. In earlier studies, the RFQ demonstrated high reliability; the Cronbach alpha coefficient was 0.70 or higher [24,25]. A special questionnaire was developed especially for this project in order to evaluate sociodemographic, work, and COVID-19-related characteristics. Based on previous studies that dealt with examining the mental health status of healthcare workers [26–29], factors such as gender (male = 1, female = 2), age, profession (doctors = 1, nurses = 2), work environment during the COVID-19 pandemic (frontline healthcare workers = 1, non-frontline healthcare workers = 2), socioeconomic status (ranging from 1 = very poor to 5 = excellent), marital status (married = 1, single = 2), and number of children (no children = 1, one child = 2, two or more children = 3) were used as the study's control variables.

2.3. Statistical Analysis

Inadequate answers, responses from participants who did not meet all the criteria needed for inclusion in the study, and responses from respondents who did not answer all the questionnaire questions were all excluded based on direct inspection prior to the statistical analysis of the data gathered. SPSS Statistics software (IBM SPSS Statistics for Windows, Version 22.0, Armonk, NY, USA) was used for statistical processing and data analysis. Measures of descriptive statistics included number (N), frequency (%), mean values, standard deviations (SD), minimum, maximum, skewness, and kurtosis. As an indicator of internal consistency, Cronbach's alpha coefficient was used to evaluate the scales' reliability. Categorical variables were transformed into numerical values for statistical purposes. Pearson's correlation coefficients, along with significance tests, were utilized to describe the relationship between the studied variables. A multiple hierarchical regression analysis was utilized to assess whether the resilience and mentalizing aspects were significant predictors of depression, anxiety, and stress as dependent variables separately.

3. Results

3.1. Characteristics of the Analyzed Sample

The sociodemographic and work-related characteristics of the respondents are shown in Table 1. Of the tested healthcare workers, the majority were females (65.8%), nurses (65.3%), and married individuals (71.7%). Of the 406 healthcare workers who participated in the study, 203 were frontline healthcare workers (64 doctors and 139 nurses), while 203 were non-frontline healthcare workers (77 doctors and 126 nurses). The doctors and nurses in this study were between the ages of 19 and 61 and 26 and 62, respectively. On a scale of 1 to 5, the majority of respondents (56.4%) gave their socioeconomic situation a score of 3, which is considered to be good.

Table 1. Characteristics of the study participants.

Characteristics		Sample Size (*N* = 406)
Age (years), mean ± SD		40.11 ± 9.41
Gender, *N* (%)	Male	139 (34.2)
	Female	267(65.8)
Profession, *N* (%)	Doctors	141 (34.7)
	Nurses	265 (65.3)
Work environment, *N* (%)	COVID-19 frontline healthcare workers	203 (50)
	Non-frontline healthcare workers	203 (50)
Marital status, *N* (%)	Married	291 (71.7)
	Single	115 (28.3)
Number of children, *N* (%)	No children	120 (29.6)
	One child	111 (27.3)
	Two or more children	175 (43.1)
Socioeconomic status, *N* (%)	Very poor	12 (3)
	Poor	71 (17.5)
	Good	229 (56.4)
	Very good	78 (19.2)
	Excellent	16 (3.9)

3.2. Measures of Descriptive Statistics of the Mentalizing, Resilience, and DASS-42 Dimensions

Table 2 shows the descriptive statistics measures and the reliability of the used scales. The results on these scales' score distribution did not differ significantly from the normal distribution, according to the values of skewness and kurtosis, which varied from -1 to 1. When depression and hypomentalizing were measured, there was only a minor deviation, which indicated that the majority of respondents had depression and hypomentalizing scores that were below average. It was expected that all of the instruments used in this study would have satisfactory, good, or excellent reliability as determined by Cronbach's alpha coefficient.

Table 2. Descriptive statistics of used measures.

Scale	Min.	Max.	Mean	SD	Skew	Kurt	α
Depression	0	42	10.63	10.63	1.04	0.26	0.96
Anxiety	0	42	10.19	9.44	0.99	0.32	0.93
Stress	0	42	17.87	11.98	0.26	-0.92	0.96
Resilience (BRS)	1.00	5.00	3.21	0.74	-0.08	0.00	0.76
Hypermentalizing (RFQ-c)	0.00	3.00	1.19	0.88	0.35	-0.96	0.82
Hypomentalizing (RFQ-u)	0.00	2.50	0.57	0.62	1.21	0.68	0.70

3.3. Correlations between the Investigated Variables

Table 3 shows the correlations between the variables. Our findings indicated that with a higher degree of resilience of the healthcare workers, their depression, anxiety, and stress were lower ($r = -0.51$, $p < 0.01$; $r = -0.51$, $p < 0.05$; $r = -0.53$, $p < 0.01$, respectively). With a greater degree of certainty in one's ability to assess intentional mental states, i.e., hypermentalizing, the experience of depression, anxiety, and stress decreased ($r = -0.42$, $p < 0.01$; $r = -0.43$, $p < 0.01$; $r = -0.49$, $p < 0.01$, respectively). With a higher level of hypomentalizing, the degree of depression, anxiety, and stress increased ($r = 0.40$, $p < 0.01$; $r = 0.46$, $p < 0.01$; $r = 0.53$, $p < 0.01$, respectively).

Table 3. Correlations between the investigated variables.

	Depression	Anxiety	Stress	BRS	RFQ-c	RFQ-u	Gender	Age	M.S.	N.C.	SE.S.	Profession
Anxiety	0.68 **											
Stress	0.69 **	0.72 **										
BRS	−0.51 **	−0.51 **	−0.53 **									
RFQ-c	−0.42 **	−0.43 **	−0.49 **	0.36 **								
RFQ-u	0.40 **	0.46 **	0.53 **	−0.27 **	−0.61 **							
Gender	0.13 **	0.25 **	0.17 **	−0.18 **	−0.11 *	0.13 **						
Age	−0.04	−0.09	−0.14 **	0.07	0.18 **	−0.15 **	0.02					
M.S.	−0.02	−0.07	0.02	0.08	0.01	0.05	−0.06	−0.18 **				
N.C.	−0.04	0.09	−0.11 *	−0.06	0.11 *	−0.06	0.12 **	0.46 **	−0.48 **			
SE.S.	−0.28 **	−0.20 **	−0.19 **	0.19 **	0.07	−0.13 **	−0.13 **	−0.10 *	0.01	−0.05		
Profession	0.10 *	0.23 **	0.10 *	−0.09	−0.06	0.12 *	0.47 **	−0.04	−0.07	0.17 **	−0.10 *	
W.E.	−0.10 *	−0.10 *	−0.13 **	0.08	0.13 **	−0.11 *	−0.15 **	0.08	−0.02	0.07	0.15 **	−0.06

Note: ** $p < 0.01$, * $p < 0.05$, BRS—brief resilience scale, RFQ-c—reflective function questionnaire certain, RFQ-u—reflective function questionnaire uncertain, M.S.—marital status, N.C.—number of children, SE.S.—socioeconomic status, W.E.—work environment (COVID-19 frontline or non-frontline).

Previous research findings showed that some control variables were important for mental health status [26–29]. Our findings suggested that females had more depression, anxiety, and stress (r = 0.13, $p < 0.01$; r = 0.25, $p < 0.01$; r = 0.17, $p < 0.01$, respectively), and at the same time, lower resilience (r = −0.18, $p < 0.01$). With the increase in the age of the respondents (r = −0.14, $p < 0.01$) and the number of children they had (r = −0.11, $p < 0.05$), the level of stress decreased. With the experience of higher socioeconomic status, respondents experienced less depression, anxiety, and stress (r = −0.28, $p < 0.01$; r = 0.20, $p < 0.01$; r = −0.19, $p < 0.01$, respectively). The findings suggested that nurses had higher levels of depression, anxiety, and stress (r = 0.10, $p < 0.05$; r = 0.20, $p < 0.01$; r = 0.10, $p < 0.05$, respectively) than doctors. Frontline healthcare workers were found to have higher levels of depression, anxiety, and stress (r = −0.10, $p < 0.05$; r = 0.10, $p < 0.05$; r = 0.13, $p < 0.01$, respectively) than non-frontline healthcare workers.

3.4. Hierarchical Linear Regression Models

After it was established that the assumptions of normality, linearity, multicollinearity, and homogeneity of variance were not violated, hierarchical linear regression analysis was applied, as shown in Table 4. There were no significant problems with multicollinearity because the variance inflation factor (VIF) for each control and predictor variable was less than 5. In the analyses of depression, anxiety, and stress, the Durbin–Watson coefficients were 2.14, 2.16, and 2.02, respectively. This demonstrated that the models did not have any significant autocorrelation problems [30].

Table 4. Hierarchical linear regression analysis of the relationship among DASS-42 dimensions, resilience, and mentalizing.

	Outcome Variable: Depression					
	Control Variables			Control Variables and Predictors		
	β	t	VIF	β	t	VIF
Gender	0.07	1.41	1.33	−0.11	−0.24	1.36
Age	−0.04	−0.73	1.33	0.06	1.45	1.37
M.S.	−0.05	−1.01	1.31	−0.02	−0.55	1.33
N.C.	−0.07	−1.28	1.69	−0.09	−1.81	1.71
Profession	0.04	0.83	1.34	0.04	0.97	1.35
SE.S.	**−0.27 **	−5.57	1.05	**−0.17 **	−4.19	1.10
W.E.	−0.03	−0.71	1.06	0.00	−0.00	1.06
BRS				**−0.37 **	−8.60	1.23
RFQ-c				**−0.16 **	−3.12	1.78
RFQ-u				**0.18 **	3.56	1.67
R^2	**0.10**			**0.39**		
adj. R^2	**0.08**			**0.37**		
F Ch.	**6.50 **			**61.96 **		

Table 4. *Cont.*

	Outcome Variable: Anxiety					
	Control Variables			Control Variables and Predictors		
	β	t	VIF	β	t	VIF
Gender	**0.12 ***	2.36	1.33	0.03	0.85	1.36
Age	**−0.14 ** **	−2.74	1.33	−0.03	−0.81	1.37
M.S.	−0.01	−0.23	1.31	0.01	0.34	1.33
N.C.	0.06	1.11	1.69	0.05	1.13	1.71
Profession	**0.10 ***	1.97	1.34	**0.10 ***	2.30	1.35
SE.S.	**−0.19 ** **	−4.10	1.05	**−0.09 ***	−2.42	1.10
W.E.	−0.03	−0.67	1.06	0.00	0.13	1.06
BRS				**−0.35 ** **	−8.40	1.23
RFQ-c				**−0.17 ** **	−3.41	1.78
RFQ-u				**0.22 ** **	4.59	1.67
R^2	**0.11**			**0.42**		
adj. R^2	**0.10**			**0.41**		
F Ch.	**7.43 ** **			**71.31 ** **		

	Outcome Variable: Stress					
	Control Variables			Control Variables and Predictors		
	β	t	VIF	β	t	VIF
Gender	**0.11 ***	2.08	1.33	0.17	0.38	1.36
Age	**−0.11 ***	−2.05	1.33	0.00	0.18	1.37
M.S.	−0.03	−0.58	1.31	−0.00	−0.10	1.33
N.C.	−0.07	−1.24	1.69	−0.09	−1.91	1.71
Profession	0.00	0.04	1.34	−0.00	−0.07	1.35
SE.S.	**−0.19 ** **	−3.94	1.05	**−0.08 ***	−2.07	1.10
W.E.	−0.05	−1.03	1.06	−0.01	−0.29	1.06
BRS				**−0.40 ** **	−9.66	1.23
RFQ-c				**−0.15 ** **	−3.15	1.78
RFQ-u				**0.26 ** **	5.44	1.67
R^2	**0.08**			**0.45**		
adj. R^2	**0.06**			**0.43**		
F Ch.	**5.12 ** **			**88.05 ** **		

Note: ** $p < 0.01$, * $p < 0.05$. F Ch.—change in F, adj. R^2—adjusted R-squared, BRS—brief resilience scale, RFQ-c—reflective function questionnaire certain, RFQ-u—reflective function questionnaire uncertain, M.S.—marital status, N.C.—number of children, SE.S.—socioeconomic status, W.E.—work environment (COVID-19 frontline or non-frontline); statistically significant correlations are bolded.

3.4.1. Depression

In the model with control variables, the value of the explained variance of the dependent variable depression was 8%; however, when control and predictor variables were taken together, the value increased to 37%, primarily because of resilience ($\beta = -0.37$, $p < 0.01$), hypomentalizing ($\beta = 0.18$, $p < 0.05$), and hypermentalizing ($\beta = -0.16$, $p < 0.05$). These findings showed that more resilience and hypermentalizing reduced depression and that more hypomentalizing increased depression. Socioeconomic status was a significant negative predictor of depression ($\beta = -0.30$, $p < 0.01$), which means that higher socioeconomic status implied less depression among healthcare workers.

3.4.2. Anxiety

Resilience ($\beta = -0.35$, $p < 0.01$) and hypermentalizing ($\beta = -0.17$, $p < 0.01$), which were statistically significant negative predictors of anxiety, increased the value of the explained variance of the dependent variable anxiety from 10% in the model with control variables to 41% when control and predictor variables were combined. The findings suggest that as an aspect of an impaired mentalizing ability, hypomentalizing increased anxiety levels ($\beta = 0.22$, $p < 0.01$). When it came to the profession variable, medical technicians/nurses had a higher level of anxiety ($\beta = 0.10$, $p < 0.05$) than doctors. Furthermore, with the increase in socioeconomic status ($\beta = -0.09$, $p < 0.05$), anxiety decreased in the respondents.

3.4.3. Stress

In the model with control variables, the value of the explained variance of the dependent variable stress was 6%, but when the control and predictor variables were combined, it increased to 43%. Resilience ($\beta = -0.40$, $p < 0.01$), hypomentalizing ($\beta = 0.26$, $p < 0.01$), and hypermentalizing ($\beta = -0.15$, $p < 0.01$) were all statistically significant predictors. The results indicated that higher levels of hypermentalizing and resilience implied lower levels of stress. The finding that hypermentalizing was an important negative predictor of stress indicated that a higher degree of confidence in one's own ability to accurately assess intentional mental states reduced the degree of experiencing stress. The finding that hypomentalizing was an important positive predictor of stress indicated that the degree of uncertainty in one's ability to accurately assess intentional mental states and the degree of stress increased together or decreased together. Along with the assessment of a higher socioeconomic status, the respondents also had less stress ($\beta = -0.08$, $p < 0.05$).

Age in the model with only control variables was a significant negative predictor of anxiety ($\beta = -0.14$, $p < 0.01$) and stress ($\beta = -0.11$, $p < 0.05$), which means that a higher number of years implied less anxiety and stress. Furthermore, in the model with only control variables, gender was a significant predictor of anxiety ($\beta = 0.12$, $p < 0.05$) and stress ($\beta = 0.11$, $p < 0.05$), which implied that women had higher levels of anxiety and stress. However, with the introduction of predictor variables into the models, age and gender ceased to be significant predictors of these two dimensions of mental health status. Marital status, number of children, and work environment were not statistically significant predictors of any of the three dimensions of mental health among healthcare workers.

4. Discussion

This research aimed to investigate the role of resilience and mentalizing capacity in anticipating reported symptoms of depression, anxiety, and stress in healthcare workers during the COVID-19 outbreak. Although research that examined the relationship between mentalizing and mental health problems in a clinical sample is numerous, there is a noticeable lack of studies in the context of work psychology and occupational health. As far as we know, the relationship between these concepts has not been previously investigated on a sample of nurses and doctors.

Given that repeated exposure to stressors can result in a variety of mental health issues, the psychological effects of COVID-19 on healthcare workers who are working during the pandemic are a significant consideration to take into account [4]. This study showed that healthcare workers during the COVID-19 pandemic in Serbia not only had higher average values of depression, anxiety, and stress than the general population in Serbia during the lockdown [3] but also more than healthcare workers in China during the peak of the COVID-19 epidemic [31], as measured and scored in the same way.

The obtained results confirmed that resilience contributed to the explanation of each of the dimensions of mental health; more precisely, the greater the resilience of healthcare workers, the lesser the depression, anxiety, and stress they experienced, which is in line with the results of numerous studies conducted around the world during the pandemic [32]. Resilience is a dynamic process of adaptability to difficult life circumstances that includes several features of personal resources and was considered a protective factor against mental health problems in healthcare workers even before the outbreak of the COVID-19 pandemic [6]. It was demonstrated that during the COVID-19 pandemic, Serbian healthcare professionals' resilience was inversely correlated with burnout, positively associated with subjective well-being, and attenuated the negative correlation between burnout and subjective well-being [6]. The research that was published as part of a large scientific project to which this study also belonged had great theoretical and practical significance because it shed light on the relationship between mentalizing, burnout syndrome, and resilience in healthcare workers during the COVID-19 pandemic [33]. According to the findings of that study, which was conducted on the same sample as the current study, resilience in healthcare workers enhances feelings of personal accomplishment at work and reduces emotional fatigue, while hypomentalizing in healthcare workers increases their emotional exhaustion and depersonalization, which, in turn, causes burnout [33]. Our findings are in line with research that showed resilience characteristics to be related to reduced levels of depressive symptoms and anxiety [34], and that resilience mediated the link between stress, anxiety symptoms, and depression during the outbreak of COVID-19 [35].

Deficits in mentalizing abilities are associated with higher prevalences of clinical levels of distress and greater emotion dysregulation [36,37]. Our findings showed that as an aspect of the impaired ability to mentalize, hypomentalizing, that is, difficulty effectively assessing the mental states that underlie behavior, increases the level of depression, anxiety, and stress among healthcare workers. Mentalizing is often simplistically understood as a synonym for the ability to empathize with other people, which is an especially important trait when it comes to professions that involve helping others. In fact, mentalizing includes a wide range of abilities that critically include the ability to see one's own behavior as coherently organized by mental states and to be psychologically distinct from others [38]. Hypomentalizing or the tendency to infer less social meaning hinders mental processes and makes it difficult to understand how harmful some actions are to others, which leads to thinking about the relevance of the ability to mentalize in the work of healthcare providers. In contrast, a good ability to mentalize involves showing empathy, reflective listening, and genuine curiosity about revealing mental states during direct communication with the interlocutor. However, hypomentalizers have a tendency to ignore objective facts about the reasons for their behavior in communication with others and to judge mental states by guessing, which can lead them to wrong conclusions [39,40]. When it comes to impaired capacity for mentalizing in healthcare workers, it can reduce their ability to understand their own behavior and that of their patients or colleagues, which leads to interpersonal miscommunications, disputes, inconveniences, depersonalization, and job dissatisfaction [33], which, in turn, can result in burnout, depression, anxiety, and stress. All this is consistent with other research showing that good mentalizing ability is a safeguard for mental health [11,33,39,40]. Therefore, the lack of consideration of mental life phenomena that influence behavior by establishing assumptions and testing them in interpersonal interaction with colleagues and patients led to experiencing more distress among our respondents.

Mentalizing enables the ability to create positive and satisfying interpersonal relationships, gives meaning to one's inner experience and the outer world, and promotes satisfying interpersonal functioning where a person manages to keep their sense of identity while feeling connected to others [41]. Scandurra et al. [42] showed that mentalizing, which is a mechanism by which individuals understand, modulate, and communicate their cognitive and emotional experiences, significantly protects against the onset of depressive and anxiety symptoms. This is in agreement with our findings, which showed that hypermentalizing, which in our respondents was understood as a higher degree of trust in one's own ability to appropriately identify intentional mental states, reduced the levels of depression, anxiety, and stress. Similar findings were obtained by Lenzo et al. from a sample of 157 bereaved participants. In their study, it was shown that anxiety was negatively correlated with hypermentalizing, i.e., certainty about mental states, and in a positive correlation with hypomentalizing, i.e., uncertainty about mental states, which indicates a type of mentalizing impairment. Depression was also negatively correlated with hypermentalizing [43]. These findings suggest that hypermentalizing has a role in preventing depression, anxiety, and stress. The hypermentalizing scale that was used both in that and in our research refers mainly to the assessment of the degree of trust in the infallibility of one's own assessment of one's own and other people's mental state [17,39]. In light of this, our research's findings indicate that healthcare professionals who have a tendency to strongly believe in the accuracy of their assessments of their own and others' mental states and who avoid verifying their assumptions about intentional mental states via direct communication with coworkers and patients, on the one hand, avoid frustrations and difficulties, which, in turn, reduces the experience of depression, anxiety, and stress. On the other hand, they continue to lack access to enough knowledge of their own and their coworkers' and patients' intentional mental states, which, as was previously established, is an aspect of hypermentalizing. In terms of sociodemographic, work, and COVID-19-related characteristics, our findings showed that age, in a model with only control variables, was a significant negative predictor of anxiety and stress, which means that older healthcare workers reported less anxiety and stress. This finding is in line with the results of research conducted by Biber et al. in the USA [44] and Yassin et al. in Jordan [45]. Moreover, in the model with only control variables, gender was a significant predictor of mental health status indicators of anxiety and stress in the sense that women had higher levels of anxiety and stress, which is in line with the studies conducted by Xiao et al. [26], Biber et al. [44], and Lai et al. [27]. However, with the introduction of predictors (resilience and mentalizing dimensions) into the models, age and gender ceased to be significant predictors of anxiety and stress.

Our findings also indicated the importance of good socioeconomic status in reducing depression, anxiety, and stress among healthcare workers, which is in line with the results of earlier research suggesting that people with lower socioeconomic status are more prone to mental health problems [46,47]. Profession was a significant predictor only in the regression model of anxiety. Nurses had higher levels of anxiety than doctors, which is consistent with previous findings from China [27].

4.1. Practical Implications

Distress during the COVID-19 pandemic is a reaction to difficult circumstances that surround all people, especially healthcare workers. Mentalizing abilities offer increased flexibility in emotional understanding, thereby helping to reduce some of the most severe effects of stress. Increasing mentalizing may serve to increase the capacity to bear the negative consequences of potentially traumatic experiences [48]. Findings from the literature suggest that resilience training can be beneficial for healthcare professionals, as resilience is considered the ability to adapt to adversity, maintain balance, maintain control, and cope with external stressors [5]. Because resilience and mental health are closely related and depend on the interaction between personal and broader social factors, such as safety and accessibility to education and employment, effective strategies to support mental health and promote resilience that focus on self-efficacy and community participation are especially important after crisis situations [49], such as a pandemic. The gold standard of care should be the development of online interventions centered on resilience and the availability of psychological support for easier stress management during the COVID-19 pandemic and dealing with the long-term consequences related to one's quality of life, personal functioning, and overall well-being [50].

Although our findings have relevant implications for clinical practice, as they emphasize the importance of mentalizing and resilience for the prevention of mental disorders in healthcare workers in the context of a chronic pandemic and increased social stressors, it should be noted that additional research is necessary to examine in more detail the factors related to COVID-19 that have an impact on the mental health of healthcare workers.

4.2. Limitations

The current study has certain limitations. The data were cross-sectional, and thus, causation cannot be sufficiently established and reverse causation cannot be ruled out. Another limitation that must be taken into account is that healthcare workers from only one tertiary care hospital were included, and thus, the results cannot be generalized to the entire workforce of healthcare professionals. Additionally, response biases, which can often be difficult to eliminate in self-reported survey research like this one, may have influenced respondents' opinions, thus limiting the results of this study.

5. Conclusions

In summary, the most important conclusions of the current study were that resilience and hypermentalizing in healthcare workers reduced depression, anxiety, and stress, and that hypomentalizing, as a failure in mentalizing characterized by low certainty about the mental state of oneself and others, increased their depression, anxiety, and stress during the COVID-19 pandemic.

Since resilience and the mentalizing capacity were shown to play very important roles when it came to preventing mental health problems, there is an urgent need to establish and implement strategies to foster resilience and enhance the capacity for mentalizing among healthcare workers in order to reduce the devastating impact of the COVID-19 pandemic on mental health.

Author Contributions: Conceptualization, T.S.; methodology, T.S., M.G., M.Z., A.K., D.D. and T.M.S.; investigation, T.S.; data curation, T.S., D.D., N.N., D.H.D., A.M. and J.D.; formal analysis, A.K., J.D., T.M.S., D.H.D., N.N., M.Z. and M.G.; validation, D.D., A.K., D.H.D., A.M., M.Z. and N.N.; writing—original draft preparation, T.S.; writing—review and editing, T.S., T.M.S., J.D. and A.M.; project administration, T.S.; visualization, T.S., M.G. and A.M.; supervision, T.S. and A.M. All authors have read and agreed to the published version of the manuscript.

Funding: This research received no external funding.

Institutional Review Board Statement: The study was conducted according to the guidelines of the Declaration of Helsinki, and approved by the Institutional Review Board of the University of Belgrade, Faculty of Philosophy, Department of Psychology, Serbia (approval number: #2021-58).

Informed Consent Statement: Informed consent was obtained from all subjects involved in the study.

Data Availability Statement: The data presented in this study are available on request from the corresponding author.

Acknowledgments: This paper was the result of a large, self-funded scientific project led by Teodora Safiye during her doctoral studies.

Conflicts of Interest: The authors declare no conflict of interest.

References

1. Wang, C.; Horby, P.W.; Hayden, F.G.; Gao, G.F. A novel coronavirus outbreak of global health concern. *Lancet* **2020**, *395*, 470–473. [CrossRef] [PubMed]
2. Serafini, G.; Parmigiani, B.; Amerio, A.; Aguglia, A.; Sher, L.; Amore, M. The psychological impact of COVID-19 on the mental health in the general population. *QJM* **2020**, *113*, 531–537. [CrossRef] [PubMed]
3. Vujčić, I.; Safiye, T.; Milikić, B.; Popović, E.; Dubljanin, D.; Dubljanin, E.; Dubljanin, J.; Čabarkapa, M. Coronavirus Disease 2019 (COVID-19) Epidemic and Mental Health Status in the General Adult Population of Serbia: A Cross-Sectional Study. *Int. J. Environ. Res. Public Health* **2021**, *18*, 1957. [CrossRef] [PubMed]
4. Kang, L.; Li, Y.; Hu, S.; Chen, M.; Yang, C.; Yang, B.X.; Wang, Y.; Hu, J.; Lai, J.; Ma, X.; et al. The mental health of medical workers in Wuhan, China dealing with the 2019 novel coronavirus. *Lancet Psychiatry* **2020**, *7*, e14. [CrossRef]
5. Cleary, M.; Visentin, D.; West, S.; Lopez, V.; Kornhaber, R. Promoting emotional intelligence and resilience in undergraduate nursing students: An integrative review. *Nurse Educ. Today* **2018**, *68*, 112–120. [CrossRef]
6. Safiye, T.; Vukčević, B.; Čabarkapa, M. Resilience as a moderator in the relationship between burnout and subjective well-being among medical workers in Serbia during the COVID-19 pandemic. *Vojnosanit. Pregl.* **2021**, *78*, 1207–1213. [CrossRef]
7. Greenberg, N.; Docherty, M.; Gnanapragasam, S.; Wessely, S. Managing mental health challenges faced by healthcare workers during COVID-19 pandemic. *BMJ* **2020**, *368*, m1211. [CrossRef] [PubMed]
8. Søvold, L.E.; Naslund, J.A.; Kousoulis, A.A.; Saxena, S.; Qoronfleh, M.W.; Grobler, C.; Münter, L. Prioritizing the Mental Health and Well-Being of Healthcare Workers: An Urgent Global Public Health Priority. *Front. Public Health* **2021**, *9*, 679397. [CrossRef]
9. Spoorthy, M.S.; Pratapa, S.K.; Mahant, S. Mental health problems faced by healthcare workers due to the COVID-19 pandemic-A review. *Asian J. Psychiatr.* **2020**, *51*, 102119. [CrossRef] [PubMed]
10. Heath, C.; Sommerfield, A.; von Ungern-Sternberg, B.S. Resilience strategies to manage psychological distress among healthcare workers during the COVID-19 pandemic: A narrative review. *Anaesthesia* **2020**, *75*, 1364–1371. [CrossRef] [PubMed]
11. Schwarzer, N.H.; Nolte, T.; Fonagy, P.; Griem, J.; Kieschke, U.; Gingelmaier, S. The relationship between global distress, mentalizing and well-being in a German teacher sample. *Curr. Psychol.* **2021**, *42*, 1239–1248. [CrossRef]

12. Wald, J.; Taylor, S.; Asmundson, G.J.; Jang, K.L.; Stapleton, J. *Literature Review of Concepts: Psychological Resiliency*; British Columbia University: Vancouver, BC, Canada, 2006.
13. Smith, B.; Dalen, J.; Wiggins, K.; Tooley, E.; Christopher, P.; Bernard, J. The Brief Resilience Scale: Assessing the Ability to Bounce Back. *Int. J. Behav. Med.* **2008**, *15*, 194–200. [CrossRef] [PubMed]
14. Herrman, H.; Stewart, D.E.; Diaz-Granados, N.; Berger, E.L.; Jackson, B.; Yuen, T. What is Resilience? *Can. J. Psychiatry* **2011**, *56*, 258–265. [CrossRef] [PubMed]
15. Allen, J.G.; Fonagy, P. Mentalizing in psychotherapy. *Textb. Psychiatry* **2014**, *6*, 1095–1118.
16. Bateman, A.; Fonagy, P. (Eds.) *Handbook of Mentalizing in Mental Health Practice*; American Psychiatric Publishing, Inc.: Arlington, VA, USA, 2019.
17. Fonagy, P.; Luyten, P.; Moulton-Perkins, A.; Lee, Y.-W.; Warren, F.; Howard, S.; Ghinai, R.; Fearon, P.; Lowyck, B. Development and Validation of a Self-Report Measure of Mentalizing: The Reflective Functioning Questionnaire. *PLoS ONE* **2016**, *11*, e0158678. [CrossRef]
18. Brugnera, A.; Zarbo, C.; Compare, A.; Talia, A.; Tasca, G.A.; De Jong, K.; Greco, A.; Greco, F.; Pievani, L.; Auteri, A.; et al. Self-reported reflective functioning mediates the association between attachment insecurity and well-being among psychotherapists. *Psychother. Res.* **2021**, *31*, 247–257. [CrossRef] [PubMed]
19. World Medical Association. World Medical Association Declaration of Helsinki: Ethical Principles for Medical Research Involving Human Subjects. *JAMA* **2013**, *310*, 2191–2194. [CrossRef] [PubMed]
20. Institute of Public Health "Dr. Milan Jovanović Batut". Latest Information about COVID-19 in the Republic of Serbia. Available online: https://covid19.rs/homepage-english/ (accessed on 22 February 2022).
21. Lovibond, S.H.; Lovibond, P.F. *Manual for the Depression Anxiety Stress Scales*, 2nd ed.; Psychology Foundation: Sydney, NSW, Australia, 1995.
22. Makara-Studzińska, M.; Tyburski, E.; Załuski, M.; Adamczyk, K.; Mesterhazy, J.; Mesterhazy, A. Confirmatory Factor Analysis of Three Versions of the Depression Anxiety Stress Scale (DASS-42, DASS-21, and DASS-12) in Polish Adults. *Front. Psychiatry* **2022**, *12*, 770532. [CrossRef]
23. Pooravari, M.; Dehghani, M.; Salehi, S.; Habibi, M. Confirmatory factor analysis of Persian version of Depression, Anxiety and Stress (DASS-42): Non-clinical sample. *Razavi Int. J. Med.* **2017**, *5*, e12021.
24. Cucchi, A.; Hampton, J.A.; Moulton-Perkins, A. Using the validated Reflective Functioning Questionnaire to investigate mentalizing in individuals presenting with eating disorders with and without self-harm. *PeerJ* **2018**, *6*, e5756. [CrossRef]
25. Handeland, T.B.; Kristiansen, V.R.; Lau, B.; Håkansson, U.; Øie, M.G. High degree of uncertain reflective functioning in mothers with substance use disorder. *Addict. Behav. Rep.* **2019**, *10*, 100193. [CrossRef]
26. Xiao, X.; Zhu, X.; Fu, S.; Hu, Y.; Li, X.; Xiao, J. Psychological impact of healthcare workers in China during COVID-19 pneumonia epidemic: A multi-center cross-sectional survey investigation. *J. Affect. Disord.* **2020**, *274*, 405–410. [CrossRef]
27. Lai, J.; Ma, S.; Wang, Y.; Cai, Z.; Hu, J.; Wei, N.; Wu, J.; Du, H.; Chen, T.; Li, R.; et al. Factors Associated With Mental Health Outcomes Among Health Care Workers Exposed to Coronavirus Disease 2019. *JAMA Netw. Open* **2020**, *3*, e203976. [CrossRef]
28. Zerbini, G.; Ebigbo, A.; Reicherts, P.; Kunz, M.; Messman, H. Psychosocial burden of healthcare professionals in times of COVID-19—A survey conducted at the University Hospital Augsburg. *Ger. Med. Sci.* **2020**, *18*, Doc05. [CrossRef]
29. Cai, Q.; Feng, H.; Huang, J.; Wang, M.; Wang, Q.; Lu, X.; Xie, Y.; Wang, X.; Liu, Z.; Hou, B.; et al. The mental health of frontline and non-frontline medical workers during the coronavirus disease 2019 (COVID-19) outbreak in China: A case-control study. *J. Affect. Disord.* **2020**, *275*, 210–215. [CrossRef] [PubMed]
30. Tabachnick, B.G.; Fidell, L.S. *Using Multivariate Statistics*, 6th ed.; Pearson: Boston, MA, USA, 2013.
31. Yang, Y.; Lu, L.; Chen, T.; Ye, S.; Kelifa, M.O.; Cao, N.; Zhang, Q.; Liang, T.; Wang, W. Healthcare Worker's Mental Health and Their Associated Predictors During the Epidemic Peak of COVID-19. *Psychol. Res. Behav. Manag.* **2021**, *14*, 221–231. [CrossRef] [PubMed]
32. Jeamjitvibool, T.; Duangchan, C.; Mousa, A.; Mahikul, W. The Association between Resilience and Psychological Distress during the COVID-19 Pandemic: A Systematic Review and Meta-Analysis. *Int. J. Environ. Res. Public Health* **2022**, *19*, 14854. [CrossRef]
33. Safiye, T.; Vukčević, B.; Gutić, M.; Milidrag, A.; Dubljanin, D.; Dubljanin, J.; Radmanović, B. Resilience, Mentalizing and Burnout Syndrome among Healthcare Workers during the COVID-19 Pandemic in Serbia. *Int. J. Environ. Res. Public Health* **2022**, *19*, 6577. [CrossRef] [PubMed]
34. Skrove, M.; Romundstad, P.; Indredavik, M.S. Resilience, lifestyle and symptoms of anxiety and depression in adolescence: The Young-HUNT study. *Soc. Psychiatry Psychiatr. Epidemiol.* **2013**, *48*, 407–416. [CrossRef] [PubMed]
35. Barzilay, R.; Moore, T.M.; Greenberg, D.M.; DiDomenico, G.E.; Brown, L.A.; White, L.K.; Gur, R.C.; Gur, R.E. Resilience, COVID-19-related stress, anxiety and depression during the pandemic in a large population enriched for healthcare providers. *Transl. Psychiatry* **2020**, *10*, 291. [CrossRef] [PubMed]
36. Greenberg, D.M.; Rudenstine, S.; Alaluf, R.; Jurist, E.L. Development and validation of the Brief-Mentalized Affectivity Scale: Evidence from cross-sectional online data and an urban community-based mental health clinic. *J. Clin. Psychol.* **2021**, *77*, 2638–2652. [CrossRef]
37. Bodnar, A.; Rybakowski, J.K. Mentalization deficit in bipolar patients during an acute depressive and manic episode: Association with cognitive functions. *Int. J. Bipolar Disord.* **2017**, *5*, 38. [CrossRef]

38. Fonagy, P.; Luyten, P. Attachment, mentalizing, and the self. In *Handbook of Personality Disorders: Theory, Research, and Treatment*; Livesley, W.J., Larstone, R., Eds.; The Guilford Press: New York, NY, USA, 2018; pp. 123–140.

39. Luyten, P.; Campbell, C.; Allison, E.; Fonagy, P. The Mentalizing Approach to Psychopathology: State of the Art and Future Directions. *Annu. Rev. Clin. Psychol.* **2020**, *16*, 297–325. [CrossRef] [PubMed]

40. Bateman, A.; Campbell, C.; Luyten, P.; Fonagy, P. A mentalization-based approach to common factors in the treatment of borderline personality disorder. *Curr. Opin. Psychol.* **2018**, *21*, 44–49. [CrossRef]

41. Freeman, C. What is mentalizing? An overview. *Br. J. Psychother.* **2016**, *32*, 189–201. [CrossRef]

42. Scandurra, C.; Dolce, P.; Vitelli, R.; Esposito, G.; Testa, R.; Balsam, K.; Bochicchio, V. Mentalizing stigma: Reflective functioning as a protective factor against depression and anxiety in transgender and gender- nonconforming people. *J. Clin. Psychol.* **2020**, *76*, 1613–1630. [CrossRef]

43. Lenzo, V.; Sardella, A.; Musetti, A.; Petralia, M.C.; Grado, I.; Quattropani, M.C. Failures in Reflective Functioning and Reported Symptoms of Anxiety and Depression in Bereaved Individuals: A Study on a Sample of Family Caregivers of Palliative Care Patients. *Int. J. Environ. Res. Public Health* **2022**, *19*, 11930. [CrossRef] [PubMed]

44. Biber, J.; Ranes, B.; Lawrence, S.; Malpani, V.; Trinh, T.T.; Cyders, A.; English, S.; Staub, C.L.; McCausland, K.L.; Kosinski, M.; et al. Mental health impact on healthcare workers due to the COVID-19 pandemic: A U.S. cross-sectional survey study. *J. Patient Rep. Outcomes* **2022**, *6*, 63. [CrossRef]

45. Yassin, A.; Al-Mistarehi, A.-H.; El-Salem, K.; Karasneh, R.A.; Al-Azzam, S.; Qarqash, A.A.; Khasawneh, A.G.; Zein Alaabdin, A.M.; Soudah, O. Prevalence Estimates and Risk Factors of Anxiety among Healthcare Workers in Jordan over One Year of the COVID-19 Pandemic: A Cross-Sectional Study. *Int. J. Environ. Res. Public Health* **2022**, *19*, 2615. [CrossRef]

46. Agberotimi, S.F.; Akinsola, O.S.; Oguntayo, R.; Olaseni, A.O. Interactions Between Socioeconomic Status and Mental Health Outcomes in the Nigerian Context Amid COVID-19 Pandemic: A Comparative Study. *Front. Psychol.* **2020**, *11*, 559819. [CrossRef]

47. Pappas, S. *How Will People React to the New Financial Crisis?* American Psychological Association: Washington, DC, USA, 2020.

48. Huang, Y.L.; Fonagy, P.; Feigenbaum, J.; Montague, P.R.; Nolte, T. London Personality and Mood Disorder Research Consortium. Multidirectional pathways between attachment, mentalizing, and posttraumatic stress symptomatology in the context of childhood trauma. *Psychopathology* **2020**, *53*, 48–58. [CrossRef] [PubMed]

49. Herrman, H. Promoting mental health and resilience after a disaster. *J. Exp. Clin. Med.* **2012**, *4*, 82–87. [CrossRef]

50. Verdolini, N.; Amoretti, S.; Montejo, L.; García-Rizo, C.; Hogg, B.; Mezquida, G.; Rabelo-da-Ponte, F.D.; Vallespir, C.; Radua, J.; Martinez-Aran, A.; et al. Resilience and mental health during the COVID-19 pandemic. *J. Affect. Disord.* **2021**, *283*, 156–164. [CrossRef] [PubMed]

International Journal of
Environmental Research and Public Health

Article

Teacher Burnout in the Time of COVID-19: Antecedents and Psychological Consequences

Anita Padmanabhanunni and Tyrone B. Pretorius *

Department of Psychology, University of the Western Cape, Bellville 7530, South Africa
* Correspondence: tpretorius@uwc.ac.za

Abstract: The important, frontline role of teachers during the COVID-19 pandemic has often gone unrecognized, and attention to their mental health and well-being is often only the focus of scholarly research. The unprecedented challenges that teachers faced during the COVID-19 pandemic and the stresses and strains associated with it have severely impacted their psychological well-being. This study examined the predictors and the psychological consequences of burnout. Participants (N = 355) were schoolteachers in South Africa who completed the Perceived Vulnerability to Disease Questionnaire, the Fear of COVID-19 Scale, the Role Orientation Questionnaire, the Maslach Burnout Inventory, the Centre for Epidemiological Depression Scale, the Beck Hopelessness Scale, the Satisfaction with Life Scale, and the trait scale of the State-Trait Anxiety Inventory. The results of a multiple regression showed that fear of COVID-19, role ambiguity, and role conflict were significant predictors of emotional exhaustion and depersonalization, while perceived infectability and role ambiguity significantly predicted personal accomplishment. Gender and age also predicted emotional exhaustion and depersonalization, respectively, and age was also a significant predictor of personal accomplishment. Generally, the dimensions of burnout were significant predictors of indices of psychological well-being—namely, depression, hopelessness, anxiety, and life satisfaction—with the exception of the association between depersonalization and life satisfaction. Our results suggest that intervention efforts to reduce burnout need to provide teachers with adequate job resources to buffer against the demands and stressors associated with their work.

Keywords: burnout antecedents; burnout consequences; psychological well-being

Citation: Padmanabhanunni, A.; Pretorius, T.B. Teacher Burnout in the Time of COVID-19: Antecedents and Psychological Consequences. *Int. J. Environ. Res. Public Health* **2023**, *20*, 4204. https://doi.org/10.3390/ijerph20054204

Academic Editor: Masahito Fushimi

Received: 29 January 2023
Revised: 23 February 2023
Accepted: 24 February 2023
Published: 27 February 2023

1. Introduction

The COVID-19 pandemic profoundly impacted the education sector in many countries. To curb the spread of the disease, governments around the world implemented severe restrictions, including the closure of all educational institutions. This necessitated a transition to emergency remote teaching, resulting in unprecedented shifts from typical instructional practices [1]. The digitization of the educational process substantially increased teachers' working hours as they needed to master the use of information communication technology, implement new pedagogical practices, and guide their students in navigating an online learning environment [2]. Teachers also had to manage the shifts in educational policies and practices that occurred during the various stages of the pandemic, contend with their own fears related to COVID-19, and manage domestic responsibilities, including caring for their own children, homeschooling, and supporting elderly family members [3].

Prior research has confirmed that teaching is a particularly stressful occupation and is associated with high rates of burnout and teacher attrition. In the United States, 46% of teachers have reported high levels of daily stress, a rate that was only matched by nurses [4]. Sources of stress for teachers include the demands of their jobs, lack of resources, lack of support from school leadership, disengaged students, discipline problems, and difficult relationships with parents [4,5]. Teacher stress has also been negatively related to job performance [6] and psychological well-being [7] and positively related to absenteeism [8] and turnover intention [7]. It has also been demonstrated that teachers' stress is negatively related to students' social adjustment and academic performance [9].

For teachers, the stressors associated with their profession, along with the demands of the pandemic, can significantly impact their mental health and lead to burnout. Job burnout is defined as a psychological syndrome that results from exposure to chronic job-related stress [10]. According to Maslach's [11] multidimensional theory of burnout, the core features of burnout are an overwhelming experience of exhaustion, a sense of cynicism and detachment from the job, appraisals of ineffectiveness, and a lack of a sense of accomplishment in the work environment. The construct of burnout consists of three dimensions: emotional exhaustion (feelings of being emotionally drained); depersonalization (an indifferent attitude toward work); and reduced personal accomplishment (negatively evaluating work-related achievements) [11]. Various studies conducted during the pandemic have highlighted increased levels of burnout among school teachers. For example, a study of German in-service teachers [2] reported elevated levels of burnout pre- and post-pandemic, particularly with regard to depersonalization and lack of personal accomplishment. Similarly, in a Spanish study, Sánchez-Pujalte and colleagues [12] found high levels of teacher emotional exhaustion during the pandemic. Female teachers were more affected by burnout compared to male teachers, while older and more experienced teachers experienced lower levels of distress. A study of Canadian teachers [13] also found increased emotional exhaustion and cynicism among school teachers. However, teachers in the sample reported a heightened sense of accomplishment as the pandemic progressed, a phenomenon attributed to experiencing a greater sense of efficacy in the management of student behavior online. Burnout can significantly impact teaching effectiveness, teachers' interactions with students and parents, teacher motivation, and teachers' ability to support their students and peers. Burnout has been found to correlate with job satisfaction [14], absenteeism [15], intention to quit [3], and job performance [15]. Several studies have also reported on the negative impact of burnout on indices of psychological well-being, including depression [16–18], anxiety [16,18], hopelessness [19,20], and suicide ideation [21,22]. Studies investigating predictors of teacher burnout have identified gender, age, self-efficacy, and institutional support as salient factors [23].

The current study uses the job demands–resources (JDR) model [24] as a lens for examining the predictors and psychological consequences of burnout among schoolteachers. The JDR model is a transactional model that has been used to understand and explain stress and burnout among schoolteachers [24]. It attributes stress and resulting burnout to a mismatch between the demands of the job and the personal (e.g., sense of self-efficacy) and organizational resources (e.g., support from managers) available to an individual [24]. Job demands refer to the features of the job that require sustained cognitive and emotional effort; job resources refer to the internal and external features of the job that facilitate the achievement of work-related tasks and reduce physical and psychological demands while also promoting personal learning and growth [24]. Job demands have the potential to contribute to role conflict and role ambiguity and can increase teachers' vulnerability to stress and adverse mental health outcomes [2]. Role conflict occurs when there are conflicting expectations in the workplace, while role ambiguity refers to uncertainty regarding the key requirements of a job and how to accomplish them [25]. Existing studies (e.g., [25]) have reported a significant association between role conflict and role ambiguity and burnout. The JDR model is sensitive to the changes in demands and resources that may occur over time. For example, some teachers who were able to negotiate the initial demands of COVID-19-related prevention measures may have found their resources depleted during subsequent waves of the pandemic, leading to emotional exhaustion and burnout.

The current study was conducted in South Africa after its initial move to online and digital education pandemic prevention measures entailed the closure of all educational institutions and the transition to emergency remote learning and teaching. However, the socioeconomic circumstances of many learners meant that many of them had no access to either technology or Wi-Fi [26]. For this reason, the government instated rotational teaching where students would attend traditional schooling on a rotational basis. A significant proportion of South African schools are located in rural or disadvantaged community settings where access to resources and facilities (e.g., running water and sanitation) are limited. In addition, overcrowded classrooms make it difficult for teachers to implement physical distancing requirements.

One study [27] examined the role of demographic variables (gender and age), COVID-19-related variables (perceived vulnerability to disease and fear of COVID-19), and role stress (role conflict and ambiguity) as potential predictors of burnout. In addition, it also examined dimensions of burnout as potential predictors of certain indices of psychological well-being—namely, depression, hopelessness, anxiety, and life satisfaction. The categorization of the variables as antecedents and consequences of burnout is presented in Table 1.

Table 1. Presumed antecedents and consequences of burnout.

Antecedents	Burnout	Consequences
Gender	Emotional exhaustion	Depression
Age	Depersonalization	Hopelessness
Perceived vulnerability to disease	Personal accomplishment	Anxiety
Perceived infectability		Life satisfaction
Germ aversion		
Role stress		
Role ambiguity		
Role conflict		

2. Materials and Methods

2.1. Participants

Participants were school teachers ($N = 355$) from across South Africa. The majority resided in the Western Cape Province (82.3%), were women (76.6%), worked in an urban area (61.7%), and taught at the primary school level (61.1%). The mean age of the sample was 41.89 (SD = 12.42), and the mean number of years in the teaching profession was 15.7 (SD = 11.75). Our sample compared favorably with population data as reported in an international survey of teaching and learning, and we found no significant differences between the demographics in our sample and those of the international survey. This international survey [28] found that 60% of teachers in South Africa are women ($\chi^2 = 0.06$, $p > 0.05$), with a mean age of 43 ($t = 1.68$, $p > 0.05$), and a mean working experience of 15 years ($t = 1.11$, $p > 0.05$). In terms of COVID-19 status, 44.5% indicated that they had not contracted the virus. A smaller proportion of teachers either suspected that they had had COVID-19 (6.8%) but had not tested for the disease, or suspected that they had the virus and confirmed this through testing (16.6%). The survey took on average 20 minutes to complete and was only available in English. However, English is a compulsory language at schooling level, and also the medium of instruction at higher education institutions where teachers receive their training.

2.2. Measures

In addition to a brief demographic survey, participants completed the following questionnaires: the Perceived Vulnerability to Disease Questionnaire (PVD-Q) [29]; the Fear of COVID-19 Scale (FCV-19S) [30]; the Role Orientation Questionnaire [31]; the Maslach Burnout Inventory (MBI) [11]; the Centre for Epidemiological Depression Scale (CES-D) [32]; the Beck Hopelessness Scale (BHS) [33]; the Satisfaction with Life Scale (SWLS) [34]; and the trait scale of the State-Trait Anxiety Inventory (STAI-T) [35].

The PVD-Q assesses beliefs about personal vulnerability to infectious diseases. Duncan and colleagues [29], in a comprehensive psychometric analysis, demonstrated that the scale consists of two conceptually distinct subscales—namely, Germ Aversion (GA) and Perceived Infectability (PI). Furthermore, these two subscales appear to have different relationships with other variables. The GA subscale (eight items) assesses emotional discomfort in circumstances associated with a high potential for disease transmission. An example item from the GA subscale is: "I prefer to wash my hands pretty soon after shaking someone's hand." The PI subscale (seven items) assesses beliefs regarding the person's susceptibility to infectious diseases. An example item from the PI subscale is: "If an illness is going around, I will get it." Responses to the 15 items are scored on a 7-point Likert scale that ranges from strongly disagree (1) to strongly agree (7). Scores on the GA subscale range between 8 and 56, while on the PI subscale they range between 7 and 49. Higher scores on the GA subscale reflect a higher level of discomfort with the potential for disease transmission. Higher scores on the PI reflect higher levels of perceived infectability. The authors of the scales provided evidence of the discriminant and convergent validity of the two subscales and reported estimates of internal consistency of 0.87 and 0.74 for the PI and GA subscales, respectively [29]. However, other studies have generally reported moderate reliability coefficients for the GA subscale (e.g., $\alpha = 0.59$ [36] and $\alpha = 0.56$ [37]).

The FCV-19S is a 7-item scale that measures emotional fear reactions toward the pandemic. Responses are scored on a 5-point Likert scale that ranges from 1 (strongly disagree) to 5 (strongly agree). The total score ranges between 7 and 35, with a higher score reflecting a higher level of fear of COVID-19. The initial validation study [30] provided evidence of concurrent validity and satisfactory internal consistency reliability ($\alpha = 0.82$). The scale has been used in a variety of contexts (e.g., Israel, Mozambique, New Zealand, and Brazil) and, generally, alpha coefficients > 0.80 have been reported [38–40]. Two South African studies have also confirmed the reliability and the unidimensional nature of the scale [41,42].

The Role Orientation Questionnaire assesses two dimensions related to perceptions of role stress—namely, role conflict (RC) and role ambiguity (RA). The RC (eight items) reflects the degree of dissonance experienced with regard to role expectations. An example item from RC is: "I have to work on unnecessary things." RA (six items) reflects lack of clarity regarding role expectations. An example item from RA is: "I know exactly what is expected of me." Responses are scored on a 6-point Likert-type scale that ranges from definitely not true of my job (1) to definitely true of my job (6). Scores on the RC subscale range between 8 and 48, while on the RA subscale they range between 6 and 36. High scores on the two scales indicate higher levels of role conflict and ambiguity. The original study [31] reported internal consistency coefficients of 0.87 and 0.82 for role ambiguity and role conflict, respectively. More recent studies have also reported satisfactory reliability coefficients (e.g., RC = 0.81, RA = 0.85 [43]; RC = 0.92, RA = 0.91 [44]).

The MBI is one of the most widely used measures of burnout. It consists of 22 items that assess three dimensions of burnout: emotional exhaustion (EE: nine items), depersonalization (DP: five items), and personal accomplishment (PA: eight items). The EE subscale is regarded as the core component of burnout and describes feelings of tiredness, fatigue, and drained emotional energy resulting from work. The DP subscale describes negative and indifferent feelings toward students and colleagues, including feelings of callousness and cynicism. The PA scale refers to sense of accomplishment and effectiveness in relation to work. Scores on the EE subscale range between 0 and 54, on the DP subscale between 0 and 30, and on the PA subscale between 0 and 48. High levels of emotional exhaustion and depersonalization as well as low levels of personal accomplishment are considered indicative of burnout. Participants respond to the 22 items on a 7-point scale ranging from never (0) to every day (6). The original study that focused on the development of the scale [11] reported satisfactory estimates of internal consistency (Cronbach's alpha) ranging between 0.69 and 0.92 and also provided evidence of convergent and discriminant validity. A review of the reliability values reported in selected studies using the MBI in educational settings indicated that the reliability of each of these studies generally ranged between 0.50 and 0.90, with the reliability of the depersonalization subscale typically being consistently lower than that of the other two subscales [45]. A previous South African study confirmed the factor structure of the MBI and provided reliability estimates ranging from

0.71 to 0.89 [46]. Maslach and colleagues [47] suggested the following cut-off scores for EE (low $\leq$ 16, moderate 17$-$26, high $\geq$ 27); DP (low $\leq$ 6, moderate 7$-$12, high $\geq$ 13); and PA (low $\leq$ 31, moderate 32$-$38, high $\geq$ 39). While recognizing the cautionary note raised by Schaufeli and Van Dierendonck [48] about using cross-national cut-off scores in relation to the MBI, the cut-off scores are merely used in the results section for illustrative purposes.

The CES-D scale assesses depression and consists of 20 symptoms. Respondents are asked to indicate how often they experienced each of the symptoms during the past week on a 4-point scale that ranges from rarely or none of the time (0) to most or all of the time (3). Scores on the CES-D range between 0 and 60. The CES-D scale has demonstrated satisfactory internal consistency (0.85–0.90) and test-retest reliability (0.51–0.67). Validity has been established through patterns of correlations with clinical ratings of depression [32]. When used with a sample of South African students, satisfactory reliability coefficients (α and ω > 0.90) for the CES-D scale have also been reported [49].

The BHS assesses the degree to which individuals' cognitive schemata are associated with pessimistic expectations. It contains 20 statements, and respondents are expected to indicate whether each statement is "true" or "false". Example items include "I do not expect to get what I really want" and "My future seems dark to me". Scores on the BHS range between 0 and 20 and higher scores indicate a greater degree of hopelessness. Internal consistency of 0.93 has been reported for the BHS, with a concurrent validity of 0.074 with clinical ratings of hopelessness and 0.60 with other scales of hopelessness [33]. The BHS has previously been used in South Africa [49], and an alpha coefficient of 0.86 was reported in that study.

The STAI-T is a 20-item measure of trait anxiety. Responses are scored on a 4-point scale that ranges from almost never (1) to almost always (4). Scores on the STAI-T range between 20 and 80. Example items include: "I worry too much over something that really doesn't matter" and "I get in a state of tension or turmoil as I think over my recent concerns and interests." The STAI-T has been used in a wide variety of contexts (e.g., Denmark, Lebanon, and China), and reliability values exceeding 0.85 have generally been reported [50–53]. In South Africa, a reliability coefficient (Cronbach's alpha) of 0.90 was reported for this scale [54].

The SWLS is the most widely used measure of life satisfaction. It consists of five items that are scored on a 7-point scale that ranges from strongly agree (7) to strongly disagree (1). Higher scores reflect higher levels of satisfaction with life. Evidence of construct, convergent, and discriminant validity, as well as satisfactory estimates of internal consistency (α > 0.75) have been reported [55]. In South Africa, Pretorius and Padmanabhanunni used the classical test theory and Mokken and Rasch's analyses to confirm the reliability, validity, and unidimensional nature of the SWLS [56].

2.3. Procedure

Google Forms was used to develop an electronic version of all the measuring instruments. Permission was obtained from Facebook administrators of groups of teachers to distribute the link during the period April to June 2021. In addition, online meetings were held with some provincial education departments to explain the purpose of the study and to invite officials to assist with the distribution of the electronic questionnaire.

2.4. Ethics

Ethical approval for the study was granted by the Humanities and Social Sciences Ethics Committee of the University of the Western Cape (ethics reference number: HS21/3/8). Participants completed the survey anonymously and provided informed consent. Participants were also provided with the authors' contact details for psychological counseling support in the event that completing the survey resulted in some distress.

2.5. Data Analysis

The IBM SPSS Statistics for Windows (version 26; IBM Corp., Armonk, NY, USA) was used for all analyses, which included descriptive statistics, reliability measures (alpha and omega), and the intercorrelations between variables. Regression analyses with gender, age, perceived vulnerability to disease, fear of COVID-19, role ambiguity, and role conflict as predictors and the dimensions of burnout as dependent variables were undertaken to determine possible antecedents of burnout. Further regression analyses with the indices of psychological well-being as the dependent variable and the dimensions of burnout as the predictors were carried out to determine the possible consequences of burnout. We examined the distribution of all scores for normality and also visually inspected all scatterplots for linearity. The indices of skewness ranged between 0.5 and -0.5 for most of the variables, reflecting that these variables had a symmetrical distribution. The exceptions were role ambiguity (skewness = 0.88), depersonalization (skewness = 0.86), and hopelessness (skewness = 0.91), indicating that, in these instances, the distributions were moderately skewed. The scatterplots confirmed the linear relationships between the variables.

3. Results

The descriptive statistics, reliability values, and intercorrelations are reported in Table 2. In terms of intercorrelations, all the dimensions of burnout were correlated with the following presumed antecedents of burnout: perceived infectability (EE: r(353) = 0.27, $p < 0.001$; DP: r(353) = 0.22, $p < 0.001$; PA: r(353) = -0.15, $p = 0.005$) and role ambiguity (EE: r(353) = 0.29, $p < 0.001$; DP: r(353) = 0.25, $p < 0.001$; PA: r(353) = -0.51, $p < 0.001$). The relationship between these variables and the dimensions of burnout were positive for emotional exhaustion and depersonalization and negative for personal accomplishment. This indicates that higher levels of perceived infectability and role ambiguity are associated with higher levels of emotional exhaustion and depersonalization and lower levels of personal accomplishment.

Table 2. Descriptive statistics, reliability values, and intercorrelations between variables.

Variable	1	2	3	4	5	6	7	8	9	10	11	12	13	14
1. Gender	—													
2. Age	0.12 *	—												
3. PI	−0.02	0.09	—											
4. Germ Aversion	0.01	0.09	0.35 **	—										
5. Fear of COVID-19	−0.09	0.05	0.41 **	0.25 **	—									
6. Role Ambiguity	−0.05	−0.07	0.13 *	−0.24 *	0.04	—								
7. Role Conflict	0.06	0.03	0.21 **	0.18 **	0.12 *	0.04	—							
8. EE	−0.14 *	−0.08	0.27 **	0.11 *	0.25 **	0.29 **	0.39 **	—						
9. DP	−0.05	−0.10	0.22 **	0.06	0.23 **	0.24 **	0.37 **	0.61 **	—					
10. PA	0.11 *	0.17 **	−0.15 **	0.15 **	−0.09	−0.51 **	−0.04	−0.33 **	−0.29 **	—				
11. Depression	−0.11 *	−0.12 *	0.33 **	0.04	0.28 **	0.38 **	0.23 **	0.53 **	0.41 **	−0.48 **	—			
12. Hopelessness	−0.13 *	0.07	0.25 **	0.01	0.25 **	0.37 **	0.22 **	0.49 **	0.36 **	−0.42 **	0.61 **	—		
13. Anxiety	−0.17 **	−0.19 **	0.38 **	0.13 *	0.33 **	0.34 **	0.27 **	0.57 **	0.39 **	−0.42 **	0.74 **	0.62 **	—	
14. Life Satisfaction	0.08	0.07	−0.16 **	0.07	−0.11 *	−0.42 **	−0.09	−0.33 **	−0.24 **	0.47 **	−0.55	−0.62 **	−0.52 **	—
Mean	—	41.9	28.7	42.9	20.9	14.7	30.4	25.0	7.5	32.0	22.0	5.7	44.9	21.9
SD	—	12.4	8.8	8.4	7.1	5.7	8.2	15.2	7.4	11.0	12.2	4.9	10.3	7.3
Minimum	–	23	7	11	7	6	8	0	0	4.80	0	0	20	5
Maximum		73	49	56	35	36	48	42	42	48	57	20	73	35
Alpha	—	—	0.78	0.65	0.91	0.83	0.83	0.94	0.85	0.84	0.92	0.89	0.91	0.90
Omega	—	—	0.78	0.66	0.91	0.83	0.83	0.94	0.86	0.84	0.93	0.89	0.91	0.90

Note. PI = perceived infectability, EE = emotional exhaustion, DP = depersonalization, PA = personal accomplishment. ** $p < 0.01$, * $p < 0.05$.

In addition to these predictor variables that all the dimensions of burnout were related to, certain dimensions of burnout were differentially related to other predictor variables: emotional exhaustion and depersonalization were positively related to fear of COVID-19 (EE: r(353) = 0.25, $p < 0.001$; DP: r(353) = 0.23, $p < 0.001$) and role conflict (EE: r(353) = 0.39, $p < 0.001$; DP: r(353) = 0.37, $p < 0.001$), thus indicating that higher levels of emotional exhaustion and depersonalization are associated with higher levels of fear of COVID-19 and role conflict. Germ aversion was positively associated with emotional exhaustion (r(353) = 0.11, $p = 0.042$). Gender was negatively associated with emotional exhaustion (r(353) = -0.14, $p = 0.024$) and positively associated with personal accomplishment (r(353) = 0.11, $p = 0.048$), indicating that women reported higher levels of emotional exhaustion and lower levels of personal accomplishment. Finally, in terms of the presumed predictor variables, age was positively associated with personal accomplishment (r(353) = 0.17, $p < 0.001$), indicating that older respondents reported higher levels of personal accomplishment.

In terms of variables presumed to be psychological consequences of burnout, emotional exhaustion and depersonalization were positively related to depression (EE: r(353) = 0.53, $p < 0.001$; DP: r(353) = 0.41, $p < 0.001$), hopelessness (EE: r(353) = 0.49, $p < 0.001$; DP: r(353) = 0.36, $p < 0.001$), and anxiety (EE: r(353) = 0.57, $p < 0.001$; DP: r(353) = 0.39, $p < 0.001$), as well as negatively related to life satisfaction (EE: r(353) = −0.033, $p < 0.001$; DP: r(353) = −0.24, $p < 0.001$). Thus, high levels of emotional exhaustion and depersonalization were associated with higher levels of depression, hopelessness, and anxiety as well as lower levels of life satisfaction. Personal accomplishment, however, was negatively related to depression (r(353) = −0.48, $p < 0.001$), hopelessness (r(353) = −0.42, $p < 0.001$), and anxiety (r(353) = −0.42, $p < 0.001$) and positively related to life satisfaction (r(353) = 0.47, $p < 0.001$). This indicates that higher levels of emotional exhaustion and depersonalization, as well as lower levels of personal accomplishment, were associated with higher levels of depression, hopelessness, and anxiety and lower levels of life satisfaction.

The mean scores for the various dimensions of burnout were as follows: EE = 25.0 (±15.2), DP = 7.5 (±7.4), and PA = 32.0 (±11). A systematic review of 94 studies reported mean scores of 20.6 for emotional exhaustion, 6.6 for depersonalization, and 28.7 for personal accomplishment [57]. The mean scores in the current study were significantly higher with respect to emotional exhaustion (t(354) = 5.44, $p < 0.001$) and depersonalization (t(354) = 2.28, $p = 0.023$), but also higher in terms of personal accomplishment (t(354) = 5.63, $p < 0.001$) when compared to those reported in the systematic review. In terms of these cut-off scores, 43.4% and 21.1% reported high and moderate emotional exhaustion, respectively; additionally, 19.2% and 23.1% reported high and moderate depersonalization, respectively, while 44.5% and 17.2% reported low and moderate personal accomplishment.

With respect to the internal consistency of the measuring instruments, the questionnaires, with the exception of the GA subscale, generally demonstrated very satisfactory reliability coefficients (α and ω = 0.78 to 0.92). The one exception was the GA subscale, which had moderate but acceptable reliability (α = 0.65; ω = 0.66).

The results of the regression analysis with the dimensions of burnout as dependent variables and demographic variables (gender and age), perceived vulnerability to disease, fear of COVID-19, role conflict, and role ambiguity as independent variables are reported in Table 3. This table reveals that the variables presumed to be antecedents of burnout were all associated with one or more dimensions of burnout, with the exception of germ aversion.

Table 3. Predicting burnout with presumed antecedents.

Predictors	Beta	SE	B	95% CI	p
Emotional Exhaustion (R^2 = 0.277)					
Gender	−2.875	1.284	−0.104	[−0.540, −0.351]	0.026
Age	−0.062	0.045	−0.065	[−0.150, 0.025]	0.163
Germ Aversion	0.032	0.074	0.023	[−0.113, 0.178]	0.663
Perceived Infectability	0.123	0.074	0.090	[−0.022, 0.268]	0.096
Fear of COVID-19	0.273	0.085	0.163	[0.105, 0.441]	0.001
Role Ambiguity	0.489	0.103	0.231	[0.287, 0.692]	0.001
Role Conflict	0.497	0.069	0.341	[0.362, 0.632]	0.001
Depersonalization (R^2 = 0.298)					
Gender	−1.852	1.08	−0.078	[−3.984, 0.280]	0.088
Age	−0.094	0.038	−0.114	[−0.169, −0.020]	0.013
Germ Aversion	0.036	0.062	0.030	[−0.086, 0.159]	0.561
Perceived Infectability	0.095	0.062	0.081	[−0.028, 0.217]	0.129
Fear of COVID-19	0.202	0.072	0.141	[0.061, 0.344]	0.005
Role Ambiguity	0.467	0.087	0.257	[0.296, 0.638]	0.001
Role Conflict	0.453	0.058	0.363	[0.339, 0.567]	0.001

Table 3. *Cont.*

Predictors	Beta	SE	B	95% CI	*p*
Personal Accomplishment ($R^2 = 0.295$)					
Gender	1.476	1.167	0.058	[−0.819, 3.771]	0.207
Age	0.120	0.041	0.136	[0.040, 0.200]	0.003
Germ Aversion	0.112	0.067	0.086	[−0.020, 0.244]	0.096
Perceived Infectability	−0.132	0.067	−0.105	[−0.264, 000]	0.050
Fear of COVID-19	−0.073	0.077	−0.047	[−0.225, 0.080]	0.348
Role Ambiguity	−0.890	0.094	−0.457	[−1.074, −0.706]	0.001
Role Conflict	−0.027	0.062	−0.020	[−0.150, 0.096]	0.665

The predictors of emotional exhaustion and depersonalization included role ambiguity (EE: $\beta = 0.231$, $p < 0.001$; DP: $\beta = 0.257$, $p < 0.001$), role conflict (EE: $\beta = 0.341$, $p < 0.001$; DP: $\beta = 0.363$, $p < 0.001$), and fear of COVID-19 (EE: $\beta = 0.163$, $p < 0.001$; DP: $\beta = 0.141$, $p = 0.005$). The predictors of personal accomplishment included role ambiguity ($\beta = -0.457$, $p < 0.001$) and perceived infectability ($\beta = -0.105$, $p = 0.05$). The only predictor of personal accomplishment and depersonalization was age (PA: $\beta = 0.136$, $p = 0.003$; DP: $\beta = -0.114$, $p = 0.013$). Finally, gender was a significant predictor of emotional exhaustion ($\beta = -0.104$, $p = 0.026$).

Predictors of psychological well-being on the basis of dimensions of burnout are reported in Table 4. All the dimensions of burnout were significant predictors of indices of psychological well-being, with the exception of depersonalization, which predicted life satisfaction.

Table 4. Dimensions of burnout as predictors of indices of psychological well-being.

Predictors	Beta	SE	B	95% CI	*p*
Depression ($R^2 = 0.389$)					
Emotional Exhaustion	0.295	0.061	0.289	[0.175, 0.414]	0.001
Depersonalization	0.200	0.072	0.168	[0.059, 0.341]	0.006
Personal Accomplishment	−0.371	0.049	−0.334	[−0.468, −0.273]	0.001
Hopelessness ($R^2 = 0.314$)					
Emotional Exhaustion	0.113	0.026	0.275	[0.062, 0.165]	0.001
Depersonalization	0.069	0.031	0.143	[0.009, 0.129]	0.025
Personal Accomplishment	−0.130	0.021	−0.290	[−0.172, −0.089]	0.001
Anxiety ($R^2 = 0.382$)					
Emotional Exhaustion	0.325	0.052	0.378	[0.224, 0.427]	0.001
Depersonalization	0.130	0.061	0.130	[0.011, 0.250]	0.032
Personal Accomplishment	−0.245	0.042	−0.261	[−0.327, −0.162]	0.001
Life Satisfaction ($R^2 = 0.251$)					
Emotional Exhaustion	−0.080	0.040	−0.132	[−0.159, −0.001]	0.047
Depersonalization	−0.053	0.047	−0.075	[−0.146, 0.040]	0.260
Personal Accomplishment	0.266	0.033	0.401	[0.201, 0.330]	0.001

Table 4 indicates that emotional exhaustion was a significant predictor of all indices of psychological well-being, including depression ($\beta = 0.289$, $p < 0.001$), hopelessness ($\beta = 0.275$, $p < 0.001$), anxiety ($\beta = 0.378$, $p < 0.001$), and life satisfaction ($\beta = -0.132$, $p = 0.047$). Similarly, personal accomplishment was a significant predictor of all indices, including depression ($\beta = -0.334$, $p < 0.001$), hopelessness ($\beta = -0.290$, $p < 0.001$), anxiety ($\beta = -0.261$, $p < 0.001$), and life satisfaction ($\beta = 0.401$, $p < 0.001$). Depersonalization predicted all the indices of psychological well-being (except life satisfaction) including depression ($\beta = 0.168$, $p = 0.006$), hopelessness ($\beta = 0.143$, $p = 0.025$), and anxiety ($\beta = 0.130$, $p = 0.032$).

4. Discussion

Burnout is a highly prevalent psychological syndrome among schoolteachers, and studies conducted during the pandemic have reported an escalation in symptoms of burnout among this population group [23]. Existing studies have confirmed that burnout is negatively associated with work engagement, job satisfaction, and physical and mental health outcomes [16]. In this study, we examined the potential role of demographic variables and COVID-19-related variables as predictors of the dimensions of burnout, as well as burnout as a predictor of psychological well-being. There were several important findings.

First, the demographic variables of gender and age were differentially related to the dimensions of burnout. Gender predicted emotional exhaustion, with women more likely to report emotional exhaustion. Existing studies have produced mixed results in terms of the role of gender in burnout. Some studies have found no gender differences in the experience of burnout [58,59], while others report inconsistent results in terms of the relationship between gender and the various dimensions of burnout. For example, Sak [60] found that men had higher emotional exhaustion and depersonalization scores than women, while men had lower scores for personal accomplishment than women. Nevertheless, other studies (e.g., [61]) have reported the same results as the current study—namely, that women report higher levels of emotional exhaustion than men (e.g., [62]). While our study is supported by a meta-analysis of 183 studies that found that women report higher levels of emotional exhaustion than men, Maslach and colleagues cautioned against a simplistic interpretation of gender differences with respect to burnout, arguing that gender differences might be the result of confounding gender with occupation (teachers are more likely to be women; soldiers are more likely to be men).

We also found that age predicted depersonalization and personal accomplishment, with older participants reporting lower levels of depersonalization and higher levels of personal accomplishment than younger participants. With respect to depersonalization, our findings contradict a meta-analysis of correlates of burnout that found that higher ages were associated with higher levels of depersonalization [63]; our findings also contradicted a study that found that older participants report higher levels of emotional exhaustion and depersonalization [64]. However, the latter study also found, in line with our results, that older participants report higher levels of personal accomplishment. This finding makes sense as older teachers are established for a longer period in their career and are, thus, likely to have accomplished more than younger teachers. However, an overview of the research conducted on burnout indicates that age differences in burnout experiences might just be an artifact of survival bias; those who experienced burnout early in their careers might have already dropped out of the profession, so it is possible that only those with low levels of burnout remain in the profession [65].

In terms of the COVID-19-related variables, fear of COVID-19 predicted emotional exhaustion and depersonalization, while perceived infectability predicted personal accomplishment. In a study of Filipino teachers during the COVID-19 pandemic, it was found that fear of COVID-19 was significantly associated with remote teaching burnout [66]. Similarly, in a sample of Egyptian physicians, it was found that fear of COVID-19 was significantly associated with all dimensions of burnout [67]; this is partly consistent with the results obtained in the current study. Experiencing fear of COVID-19 is a stressful state, and the relationship between stress and burnout is well-established in the research literature (e.g., [68,69]). It is likely that the same mechanism underlies perceived infectability; in other words, perceiving oneself as more susceptible to infections may lead to increased stress, in turn leading to burnout. In the South African context, it is likely that teachers experienced a heightened fear of COVID-19 owing to the resumption of traditional schooling and difficulties implementing personal protective measures in schools. Overcrowded classrooms, lack of access to clean running water, and limited personal protective equipment could have contributed to teachers fearing the possibility of contagion.

Role stress, in the form of role ambiguity and role conflict, was a significant predictor of all the dimensions of burnout with the exception of role conflict, which was not a significant predictor of personal accomplishment. These results are consistent with findings in other studies regarding the relationship between role stress and burnout (e.g., [70,71]). In the context of this pandemic, during which teachers rapidly had to adjust to new ways of working, the potential for role conflict and role ambiguity is significant. Being confused about the new parameters of one's role (role ambiguity) and the irreconcilable demands of wanting to deliver effective teaching while doing so either remotely or on a rotational basis (role conflict) increases stress levels, which leads to burnout.

In terms of the relationship between burnout and indices of psychological well-being, all the dimensions of burnout significantly predicted depression, hopelessness, anxiety, and life satisfaction, with the exception of depersonalization, which was not a significant predictor of life satisfaction. These findings confirm the well-established association between burnout and psychological well-being (e.g., [72,73]).

The findings of this study have shown that multiple factors could lead to burnout, which is, in turn, associated with negative psychological well-being. As indicated, teaching is a highly stressful occupation with an increased risk of burnout. This necessitates investment in programs and interventions to help teachers cope with their stressful work environments. Maslach and colleagues [65] indicated that interventions aimed at burnout reduction have either focused on the individual or on the organization. At an individual level, this could include psycho-educational approaches or evidence-based therapies aimed at teaching people to cope with stress. At an organizational level, this could include support from school leadership, workplace wellness promotion programs, and mentoring programs. School leadership could also receive training to monitor the early warning signs of burnout and, thus, identify teachers at risk of burnout. Finally, authorities should do more to ensure that the working environment of teachers is conducive to quality learning and teaching.

The study has certain limitations. The cross-sectional nature of the data precludes the drawing of any causal inferences. Participation was voluntary, potentially leading to selection bias. Additionally, self-report measures, which are vulnerable to social desirability effects, were used. The majority of the sample were female and from one geographic area. In future studies, a more diverse sample could further corroborate our results.

5. Conclusions

Teachers are among a country's most valuable resources. Teaching, however, is one of the most stressful occupations, and the mental health of teachers is often neglected. This could potentially lead to increased disillusionment with the teaching profession, high turnover, and negative student learning outcomes. To better understand burnout syndrome in teachers, this study examined multiple factors associated with burnout as well as the relationship between burnout and psychological well-being. We found that demographic variables, COVID-19-related variables, and role stress were significantly associated with the dimensions of burnout. The dimensions of burnout also predicted depression, hopelessness, anxiety, and life satisfaction. This study underscores the need for interventions to improve the working conditions of teachers and actively implement programs aimed at reducing burnout.

Author Contributions: Conceptualization, A.P. and T.B.P.; Formal analysis, T.B.P.; Investigation, A.P. and T.B.P.; Methodology, T.B.P.; Project administration, A.P.; Supervision, A.P.; Visualization, T.B.P.; Writing—original draft, A.P. and T.B.P.; Writing—review and editing, A.P. and T.B.P. All authors have read and agreed to the published version of the manuscript.

Funding: This research received no external funding.

Institutional Review Board Statement: The study was conducted according to the guidelines of the Declaration of Helsinki and approved by the Humanities and Social Science Research Ethics Committee of the University of the Western Cape (Ethics reference number HS21/3/8 14 May 2021).

Informed Consent Statement: Informed consent was obtained from all subjects involved in the study.

Data Availability Statement: The raw data supporting the conclusions of this article will be made available by the authors, without undue reservation.

Conflicts of Interest: The authors declare no conflict of interest.

References

1. Li, B.; Li, Z.; Fu, M. Understanding beginning teachers' professional identity changes through job demands-resources theory. *Acta Psychol.* **2022**, *230*, 103760. [CrossRef] [PubMed]
2. Weißenfels, M.; Klopp, E.; Perels, F. Changes in Teacher Burnout and Self-Efficacy During the COVID-19 Pandemic: Interrelations and e-Learning Variables Related to Change. *Front. Educ.* **2022**, *6*, 736992. [CrossRef]
3. Madigan, D.J.; Kim, L.E. Towards an understanding of teacher attrition: A meta-analysis of burnout, job satisfaction, and teachers' intentions to quit. *Teach. Teach. Educ.* **2021**, *105*, 103425. [CrossRef]
4. Greenberg, M.; Brown, J.; Abenavoli, R. Teacher stress and health. Social and Emotional Learning. *RWJF Collect.* **2016**, *9*, 1–12.
5. Humantold. Why Being a Teacher Is Stressful. 2021. Available online: https://humantold.com/blog/why-being-a-teacher-is-stressful/ (accessed on 13 January 2023).
6. Yunarti, B.S.; Asaloei, S.I.; Wula, P.; Werang, B.R. Stress and performance of elementary school teachers of Southern Papua: A survey approach. *Univers. J. Educ. Res.* **2020**, *8*, 924–930. [CrossRef]
7. Nazari, M.; Alizadeh Oghyanous, P. Exploring the role of experience in L2 teachers' turnover intentions/occupational stress and psychological well-being/grit: A mixed methods study. *Cogent Educ.* **2021**, *8*, 1892943. [CrossRef]
8. Howard, J.T.; Howard, K.J. The effect of perceived stress on absenteeism and presenteeism in public school teachers. *J. Workplace Behav. Health* **2020**, *35*, 100–116. [CrossRef]
9. Hoglund, W.L.G.; Klingle, K.E.; Hosan, N.E. Classroom risks and resources: Teacher burnout, classroom quality and children's adjustment in high needs elementary schools. *J. Sch. Psychol.* **2015**, *53*, 337–357. [CrossRef]
10. Maslach, C.; Leiter, M.P. Understanding the burnout experience: Recent research and its implications for psychiatry. *World Psychiatry* **2016**, *15*, 103–111. [CrossRef]
11. Maslach, C.; Jackson, S.E. The Measurement of Experienced Burnout. *J. Occup. Behav.* **1981**, *2*, 99–113. [CrossRef]
12. Sánchez-Pujalte, L.; Mateu, D.N.; Etchezahar, E.; Gómez Yepes, T. Teachers' Burnout during COVID-19 Pandemic in Spain: Trait Emotional Intelligence and Socioemotional Competencies. *Sustainability* **2021**, *13*, 7259. [CrossRef]
13. Sokal, L.; Trudel, L.E.; Babb, J. Canadian teachers' attitudes toward change, efficacy, and burnout during the COVID-19 pandemic. *Int. J. Educ. Res.* **2020**, *1*, 100016. [CrossRef]
14. Khamisa, N.; Peltzer, K.; Ilic, D.; Oldenburg, B. Work related stress, burnout, job satisfaction and general health of nurses: A follow-up study. *Int. J. Nurs. Pract.* **2016**, *22*, 538–545. [CrossRef] [PubMed]
15. Dyrbye, L.N.; Shanafelt, T.D.; Johnson, P.O.; Johnson, L.A.; Satele, D.; West, C.P. A cross-sectional study exploring the relationship between burnout, absenteeism, and job performance among American nurses. *BMC Nurs.* **2019**, *18*, 57. [CrossRef]
16. Koutsimani, P.; Montgomery, A.; Georganta, K. The Relationship between Burnout, Depression, and Anxiety: A Systematic Review and Meta-Analysis. *Front. Psychol.* **2019**, *10*, 284. [CrossRef] [PubMed]
17. Meier, S.T.; Kim, S. Meta-regression analyses of relationships between burnout and depression with sampling and measurement methodological moderators. *J. Occup. Health Psychol.* **2022**, *27*, 195–206. [CrossRef] [PubMed]
18. Fischer, R.; Mattos, P.; Teixeira, C.; Ganzerla, D.S.; Rosa, R.G.; Bozza, F.A. Association of Burnout with Depression and Anxiety in Critical Care Clinicians in Brazil. *JAMA Netw. Open* **2020**, *3*, e2030898. [CrossRef]
19. Civilotti, C.; Acquadro Maran, D.; Garbarino, S.; Magnavita, N. Hopelessness in Police Officers and Its Association with Depression and Burnout: A Pilot Study. *Int. J. Environ. Res. Public Health* **2022**, *19*, 5169. [CrossRef]
20. Lee, H.-S.; Bae, S.-Y. Analysis of convergent influence of academic burnout, anxiety and psychosocial stress on hopelessness among some health college students using structural equation model. *J. Korea Converg. Soc.* **2019**, *10*, 165–172. [CrossRef]
21. Menon, N.K.; Shanafelt, T.D.; Sinsky, C.A.; Linzer, M.; Carlasare, L.; Brady, K.J.S.; Stillman, M.J.; Trockel, M.T. Association of Physician Burnout with Suicidal Ideation and Medical Errors. *JAMA Netw. Open* **2020**, *3*, e2028780. [CrossRef]
22. Lheureux, F.; Truchot, D.; Borteyrou, X. Suicidal tendency, physical health problems and addictive behaviours among general practitioners: Their relationship with burnout. *Work Stress* **2016**, *30*, 173–192. [CrossRef]
23. Gillet, N.; Morin, A.J.S.; Sandrin, É.; Fernet, C. Predictors and outcomes of teachers' burnout trajectories over a seven-year period. *Teach. Teach. Educ.* **2022**, *117*, 103781. [CrossRef]
24. Bakker, A.B.; Demerouti, E. The Job Demands-Resources model: State of the art. *J. Manag. Psychol.* **2007**, *22*, 309–328. [CrossRef]
25. Xu, L. Teacher-researcher role conflict and burnout among Chinese university teachers: A job demand-resources model perspective. *Stud. High. Educ.* **2019**, *44*, 903–919. [CrossRef]
26. Maree, J.G. Managing the COVID-19 pandemic in South African Schools: Turning challenge into opportunity. *S. Afr. J. Psychol.* **2022**, *52*, 249–261. [CrossRef]
27. Sithole, S. *School Functionality in the COVID-19 Environment: Practical Examples of What Works and Does Not Work*; Department of Basic Education: Pretoria, South Africa, 2021.
28. OECD. Results from Talis 2019. 2019. Available online: https://www.oecd.org/education/talis/ (accessed on 24 August 2022).

29. Duncan, L.A.; Schaller, M.; Park, J.H. Perceived vulnerability to disease: Development and validation of a 15-item self-report instrument. *Pers. Individ. Differ* **2009**, *47*, 541–546. [CrossRef]

30. Ahorsu, D.K.; Lin, C.Y.; Imani, V.; Saffari, M.; Griffiths, M.D.; Pakpour, A.H. The fear of COVID-19 scale: Development and initial validation. *Int. J. Ment. Health Addict.* **2020**, *20*, 1537–1545. [CrossRef]

31. Rizzo, J.R.; House, R.J.; Lirtzman, S.I. Role Conflict and Ambiguity in Complex Organizations. *Adm. Sci. Q.* **1970**, *15*, 150–163. [CrossRef]

32. Radloff, L.S. The CES-D scale: A self-report depression scale for research in the general population. *Appl. Psychol. Meas.* **1977**, *1*, 385–401. [CrossRef]

33. Beck, A.T.; Weissman, A.; Lester, D.; Trexler, L. The measurement of pessimism: The hopelessness scale. *J. Consult. Clin. Psychol.* **1974**, *42*, 861–865. [CrossRef]

34. Diener, E.D.; Emmons, R.A.; Larsen, R.J.; Griffin, S. The satisfaction with life scale. *J. Person. Assess.* **1985**, *49*, 71–75. [CrossRef] [PubMed]

35. Spielberger, C.D. *Manual for the State-Trait Anxiety Inventory*; Consulting Psychologists Press: Palo Alto, CA, USA, 1983.

36. Díaz, A.; Beleña, Á.; Zueco, J. The role of age and gender in perceived vulnerability to infectious diseases. *Int. J. Environ. Res. Public Health* **2020**, *17*, 485. [CrossRef] [PubMed]

37. Prokop, P.; Fancovicova, J. The Effect of Owning Animals on Perceived Vulnerability to, and Avoidance of, Parasitic Diseases in Humans. *J. Individ. Differ.* **2011**, *32*, 129–136. [CrossRef]

38. Bitan, T.D.; Grossman-Giron, A.; Bloch, Y.; Mayer, Y.; Shiffman, N.; Mendlovic, S. Fear of COVID-19 scale: Psychometric characteristics, reliability and validity in the Israeli population. *Psychiatry Res.* **2020**, *289*, 113100. [CrossRef]

39. Giordani, R.C.F.; Zanoni da Silva, M.; Muhl, C.; Giolo, S.R. Fear of COVID-19 scale: Assessing fear of the coronavirus pandemic in Brazil. *J. Health Psychol.* **2022**, *27*, 901–912. [CrossRef]

40. Winter, T.; Riordan, B.C.; Pakpour, A.H.; Griffiths, M.D.; Mason, A.; Poulgrain, J.W.; Scarf, D. Evaluation of the English version of the Fear of COVID-19 Scale and its relationship with behavior change and political beliefs. *Int. J. Ment. Health Addict.* **2023**, *21*, 372–382. [CrossRef]

41. Makhubela, M.; Mashegoane, S. Psychometric properties of the Fear of COVID-19 Scale amongst black South African university students. *Afr. J. Psychol. Assess.* **2021**, *3*, e1–e6. [CrossRef]

42. Pretorius, T.B.; Padmanabhanunni, A.; Stiegler, N.; Bouchard, J.-P. Validation of the fear of COVID-19 scale in South Africa: Three complementary analyses. *Ann. Med. Psychol.* **2021**, *179*, 940. [CrossRef]

43. Haji Mohammad Hoseini, M.; Asayesh, H.; Amaniyan, S.; Sharififard, F.; Elahi, A.; Yaghoubi Kopaie, S. Role conflict and role ambiguity as predictors of turnover intention among nurses. *J. Nurs. Midwifery Sci.* **2021**, *8*, 49–53. [CrossRef]

44. Üngüren, E.; Arslan, S. The effect of role ambiguity and role conflict on job performance in the hotel industry: The mediating effect of job satisfaction. *Tour. Manag.* **2021**, *17*, 45–58. [CrossRef]

45. Hawrot, A.; Koniewski, M. Factor Structure of the Maslach Burnout Inventory–Educators Survey in a Polish-Speaking Sample. *J. Career Assess.* **2018**, *26*, 515–530. [CrossRef]

46. Pretorius, T. Using the Maslach Burnout Inventory to assess educator's burnout at a university in South Africa. *Psychol. Rep.* **1994**, *75*, 771–777.

47. Maslach, C.; Jackson, S.E.; Leiter, M.P. *Maslach Burnout Inventory Manual*; Consulting Psychologists Press: Palo Alto, CA, USA, 1996.

48. Schaufeli, W.B.; Van Dierendonck, D. A cautionary note about the cross-national and clinical validity of cut-off points for the Maslach Burnout Inventory. *Psychol. Rep.* **1995**, *76*, 1083–1090. [CrossRef] [PubMed]

49. Padmanabhanunni, A.; Pretorius, T.B. When coping resources fail: The health-sustaining and moderating role of fortitude in the relationship between COVID-19-related worries and psychological distress. *Afr. Saf. Promot. J. Inj. Violence Prev.* **2020**, *18*, 28–47.

50. Gustafson, L.W.; Gabel, P.; Hammer, A.; Lauridsen, H.H.; Petersen, L.K.; Andersen, B.; Bor, P.; Larsen, M.B. Validity and reliability of State-Trait Anxiety Inventory in Danish women aged 45 years and older with abnormal cervical screening results. *BMC Med. Res. Methodol.* **2020**, *20*, 89. [CrossRef]

51. Hallit, S.; Haddad, C.; Hallit, R.; Akel, M.; Obeid, S.; Haddad, G.; Soufia, M.; Khansa, W.; Khoury, R.; Kheir, N.; et al. Validation of the Hamilton Anxiety Rating Scale and State Trait Anxiety Inventory A and B in Arabic among the Lebanese population. *Clin. Epidemiol. Glob. Health* **2020**, *8*, 1104–1109. [CrossRef]

52. Han, J.; Yin, H.; Wang, J.; Zhang, J. Job demands and resources as antecedents of university teachers' exhaustion, engagement and job satisfaction. *Educ. Psychol.* **2020**, *40*, 318–335. [CrossRef]

53. Han, Y.; Fan, J.; Wang, X.; Xia, J.; Liu, X.; Zhou, H.; Zhang, Y.; Zhu, X. Factor Structure and Gender Invariance of Chinese Version State-Trait Anxiety Inventory (Form Y) in University Students. *Front. Psychol.* **2020**, *11*, 2228. [CrossRef]

54. Pretorius, T.; Padmanabhanunni, A. A looming mental health pandemic in the time of COVID-19? Role of fortitude in the interrelationship between loneliness, anxiety, and life satisfaction among young adults. *S. Afr. J. Psychol.* **2021**, *51*, 256–268. [CrossRef]

55. Pavot, W.; Diener, E. The Satisfaction with Life Scale and the emerging construct of life satisfaction. *J. Posit. Psychol.* **2008**, *3*, 137–152. [CrossRef]

56. Pretorius, T.B.; Padmanabhanunni, A. Assessing the Cognitive Component of Subjective Well-Being: Revisiting the Satisfaction with Life Scale with Classical Test Theory and Item Response Theory. *Afr. J. Psychol. Assess.* **2022**, *4*, 1–9. [CrossRef]

57. Squillaci Lanners, M. Are teachers more affected by burnout than physicians, nurses and other professionals? A systematic review of the literature. In Proceedings of the International Conference on Applied Human Factors and Ergonomics, Washington, DC, USA, 24–28 July 2019; pp. 147–155.
58. Russell, D.W.; Altmaier, E.; Van Velzen, D. Job-Related Stress, Social Support, and Burnout among Classroom Teachers. *J. Appl. Psychol.* **1987**, *72*, 269–274. [CrossRef] [PubMed]
59. Kroupis, I.; Kourtessis, T.; Kouli, O.; Tzetzis, G.; Derri, V.; Mavrommatis, G. Job satisfaction and burnout among Greek P.E. teachers. A comparison of educational sectors, level and gender. (Satisfacción laboral y burnout de los profesores de la educación física en Grecia). *Cult. Cienc. Deporte* **2017**, *12*, 5–14. [CrossRef]
60. Sak, R. Gender Differences in Turkish Early Childhood Teachers' Job Satisfaction, Job Burnout and Organizational Cynicism. *Early Child. Educ. J.* **2018**, *46*, 643–653. [CrossRef]
61. Redondo-Flórez, L.; Tornero-Aguilera, J.F.; Ramos-Campo, D.J.; Clemente-Suárez, V.J. Gender Differences in Stress- and Burnout-Related Factors of University Professors. *BioMed Res. Int.* **2020**, *2020*, 6687358–6687359. [CrossRef]
62. León-Rubio, J.M.; León-Pérez, J.M.; Cantero, F.J. Prevalencia y factores predictivos del burnout en docentes de la enseñanza pública: El papel del género. *Ansiedad Estrés* **2013**, *19*, 11–25.
63. Park, E.-Y.; Shin, M. A Meta-Analysis of Special Education Teachers' Burnout. *SAGE Open* **2020**, *10*, 215824402091829. [CrossRef]
64. Rostami, S.; Ghanizadeh, A.; Ghapanchi, Z. A study of contextual precursors of burnout among EFL teachers. *Int. J. Res. Stud. Psychol.* **2015**, *4*, 13–24. [CrossRef]
65. Maslach, C.; Schaufeli, W.B.; Leiter, M.P. Job burnout. *Annu. Rev. Psychol.* **2001**, *52*, 397–422. [CrossRef]
66. Carreon, T.; Rotas, E.; Cahapay, M.; Garcia, K.; Amador, R.; Anoba, J.L. Fear of COVID-19 and Remote Teaching Burnout of Filipino K to 12 Teachers. *Int. J. Educ. Res. Innov.* **2021**, 552–567. [CrossRef]
67. Abdelghani, M.; El-Gohary, H.M.; Fouad, E.; Hassan, M.S. Addressing the relationship between perceived fear of COVID-19 virus infection and emergence of burnout symptoms in a sample of Egyptian physicians during COVID-19 pandemic: A cross-sectional study. *Middle East Curr. Psychiatry* **2020**, *27*, 70. [CrossRef]
68. Bottiani, J.H.; Duran, C.A.K.; Pas, E.T.; Bradshaw, C.P. Teacher stress and burnout in urban middle schools: Associations with job demands, resources, and effective classroom practices. *J. Sch. Psychol.* **2019**, *77*, 36–51. [CrossRef] [PubMed]
69. Vargas Rubilar, N.; Oros, L.B. Stress and Burnout in Teachers During Times of Pandemic. *Front. Psychol.* **2021**, *12*, 756007. [CrossRef] [PubMed]
70. Tang, X.; Li, X. Role Stress, Burnout, and Workplace Support Among Newly Recruited Social Workers. *Res. Soc. Work Pract.* **2021**, *31*, 529–540. [CrossRef]
71. Wu, G.; Hu, Z.; Zheng, J. Role Stress, Job Burnout, and Job Performance in Construction Project Managers: The Moderating Role of Career Calling. *Int. J. Environ. Res. Public Health* **2019**, *16*, 2394. [CrossRef]
72. Wei, H.; Dorn, A.; Hutto, H.; Webb Corbett, R.; Haberstroh, A.; Larson, K. Impacts of Nursing Student Burnout on Psychological Well-Being and Academic Achievement. *J. Nurs. Educ.* **2021**, *60*, 369–376. [CrossRef] [PubMed]
73. Rehman, A.U.; Bhuttah, T.M.; You, X. Linking Burnout to Psychological Well-being: The Mediating Role of Social Support and Learning Motivation. *Psychol. Res. Behav. Manag.* **2020**, *13*, 545–554. [CrossRef]

International Journal of
Environmental Research and Public Health

MDPI

Article

An Exploratory Study of Nurses' Feelings about COVID-19 after Experiencing SARS

Hui-Ling Lee [1], Pei-Ju Chang [2] and Li-Chiu Lin [3,*]

[1] Department of Nursing, University of Kang Ning, Taipei 114311, Taiwan
[2] Chang Hua Hospital, MOHW, Changhua 513007, Taiwan
[3] Nursing Department, Hong Kuang University, Taichung City 433304, Taiwan
* Correspondence: chiulilin@hotmail.com; Tel.: +886-934-065-608; Fax: +886-4-2633-1198

Abstract: The outbreak of severe acute respiratory syndrome (SARS) in 2003 in Taiwan impacted Taiwanese society. However, the first case of coronavirus disease 2019 (COVID-19) was reported in Wuhan and spread around the world. During these outbreaks, nursing staff experienced different levels of pressure. Studies have explored the stress and adjustment of nurses during these periods, but studies describing the feelings of nurses during both SARS and COVID-19 outbreaks are lacking. The aim of this study was to explore the experiences of nurses who had cared for both SARS and COVID-19 patients. A qualitative study combined with snowball sampling was applied. Semi-structured questions were used to interview 10 nurses who had experienced both SARS and COVID-19. Two themes and four sub-themes were analyzed, which were: facing the epidemic from the unknown to known; and the experiences from ignorance to proficiency. The sub-themes were: the feeling of frustration and concern; bottlenecks and pressures in my work; my mission and support; and positive energy and camaraderie. The results showed that the media acts as an important resource during disease outbreaks; therefore, government departments have to use their wisdom to make good use of the media. Secondly, understanding the general public's response to the disease is also important for first-line nurses. Finally, on-the-job education and guidelines for first-line nurses are necessary, and support from the administration is also important.

Keywords: SARS; COVID-19; experience; epidemic; stress

check for updates

Citation: Lee, H.-L.; Chang, P.-J.; Lin, L.-C. An Exploratory Study of Nurses' Feelings about COVID-19 after Experiencing SARS. *Int. J. Environ. Res. Public Health* **2023**, *20*, 2256. https://doi.org/10.3390/ijerph20032256

Academic Editor: Masahito Fushimi

Received: 31 December 2022
Revised: 21 January 2023
Accepted: 25 January 2023
Published: 27 January 2023

1. Introduction

In 2002, an outbreak of severe acute respiratory syndrome (SARS) occurred suddenly. It caused a nosocomial infection in Taipei City Hospital, Heping Fuyou Branch. This incident had a significant impact on all medical staff in Taiwan at that time [1,2]. However, during that time, the media in Taiwan failed to fulfill its responsibility to assist the government in advocating policies [3] and tended to exaggerate and sensationalize the reports on SARS, that caused panic and anxiety among people [3].

Although the entire outbreak in Taiwan lasted only half a year, this period also resulted in the death of 11 medical staff [4]. Such a situation caused a wave of resignations of nursing staff. For the nursing staff who remained in the workplace, the situation caused a lot of emotional trauma and exhaustion [5]. As time goes by, the SARS epidemic is slowly forgotten in the hearts of the people.

However, in December 2019, a severe infectious pneumonia, coronavirus disease 2019 (COVID-19), suddenly broke out and spread all over the world. The global epidemic of COVID-19 is quite serious and has not yet eased. The media is constantly reporting on foreign epidemics, and due to the continuous mutation of the virus many countries have begun to implement city closure measures, with medical resources being almost exhausted, causing some nurses to be under intense pressure [6]. As of 24 December 2022, the number of confirmed cases worldwide was 651,960,325, and the global fatality rate was 1.02 % [7].

In Taiwan, because of the lessons learned from SARS in 2002, there was no large-scale infection in the early stage of the epidemic and most of the confirmed cases were immigrants, due to the gradual opening of the border for entry and the gradual relaxation of epidemic prevention policies. In May 2021 and May 2022, community cluster infections in Taiwan were also caught in the epidemic storm [8].

Compared with SARS, nurses facing COVID-19 have better equipment and related training, and especially nurses who have experienced SARS, have better coping methods for COVID-19. However, due to the long epidemic of COVID-19 and the death rate being higher than that of SARS, no country in the world has been spared. Social and family support, clear health policies, and relative benefits, are very important to first-line nursing staff.

2. Literature Review

During the outbreak of SARS 18 years ago, there were clusters of infections and Taipei City Hospital, Heping Fuyou Branch, decided to recall all personnel in the hospital for isolation. Because of the emergency closure of the hospital, relevant supporting measures, protective equipment, and the control of isolation moving lines were not fully established [1,2]. All medical staff, patients, and family members were exposed to high-risk cross-infection environments, causing them to have nowhere to seek help in the hospital, and they were deeply afraid of getting sick and dying [1,2]. The closure of the hospital, coupled with unclear policies, epidemic prevention measures adjusted at any time, and insufficient protective equipment, led to negative emotions and pressure on the nursing staff [9,10].

However, the working condition was not the only pressure to nurses, because of the media. As media had been called the "forth Estate", it is powerful in affecting the public and has its social responsibilities to take [11]. Nevertheless, during the epidemic of SARS, the media mainly emphasized the "impact" dimension, with less concern in health education [11]. Lee et al. [12] (p. 353) also stated that in the period of SARS, the media coverage of the new of epidemic also reported statements which included: "nurses tried to escape from quarantined hospital", "hospital makes are in short supply make" or "disciplinary actions upon the hospitals failing to report SARS cases", which caused nurses extreme stress and they experienced significant psychological conflict between their duties as a nurse, as well as their concern of their safety.

Because of the media, the first-line nurses were labeled, which indirectly affected the life of their families in the community, as well as their children's learning at school [9,13]. Fortunately, the first SARS case was discovered in Taiwan on 14 March 2003. On 5 July 2003, the World Health Organization announced that Taiwan would be removed from the list of SARS-infected areas. The epidemic ended within four months [14].

The epidemic has been over for many years. Everyone's work and life had returned to normal, but, at the end of 2019, COVID-19 began to spread. Due to the experiences learned from SARS, Taiwan was able to strictly control the epidemic in its early stages. However, hospitals and the public in Taiwan were also affected by COVID-19 [15]. COVID-19 caused clusters in communities in Taiwan, which transmitted to hospitals. The government and administration in hospitals decided to clear the hospitals instead of closing them. Once community infection spreads to the medical system, the biggest worry is the domino effect of hospital collapse, so the burdens of medical care have to be reduced [7,15]. However, COVID-19 has been going on for three years and has not stopped; it has caused a significant impact on society around the work needed [15].

Although in the face of COVID-19, with no end in sight, some studies have pointed out that nurses who have experienced SARS have made better adjustments in the face of COVID-19 [5]. Nurses who receive on-the-job education are able to reduce their fear [5], and, after adapting to the environment, negative emotions gradually reduce [16]. At the same time, a study by Fang et al. [1] also found that nurses who have experience in caring for patients with infectious diseases have lower psychological distress and better stress relief. However, a study by Huang et al. [13] found that when nurses who had experienced

SARS had considerable knowledge of caring for the disease and sufficient equipment their stress and considerations shifted to matters related to themselves, such as worrying about whether they would infect their own family members.

Compared with SARS, nurses facing COVID-19 have better equipment and related training, and especially nurses who have experienced SARS, have better coping methods for COVID-19. However, due to the long epidemic of COVID-19 and the death rate being higher than that of SARS, no country in the world has been spared. Social and family support, as well as clear policies and relative benefits, are particularly important to first-line nursing staff.

3. Method

3.1. Research Design

In order to obtain the deep experience of the participants, this study adopted qualitative research to collect data in the form of in-depth interviews and semi-structured interview guidelines. The researchers used verbal and non-verbal communication modes to explore the feelings and responses of nurses who had cared for SARS patients prior to caring for COVID-19 patients.

3.2. Participants

A total of 10 participants were interviewed. The main research objects of this study were nurses who had experienced both SARS and COVID-19. Because the cases were not easy to obtain, the cases were collected by snowballing. We accepted cases until the data were saturated. The inclusion criteria were: (1) nursing staff who had directly cared for SARS and COVID-19 cases; and (2) nursing staff who could speak clearly and were willing to participate in the research. The basic information of the participants is shown in Table 1.

Table 1. Demographic data.

No.	Age	Sex	Education	Marriage	Position		Total Seniority (Years)
					SARS	**COVID-19**	
A	43	F	College	Married	Nurse, Duty ward	Nurse, Isolation ward	22
B	55	F	University	Married	Nurse, ER	Nurse, OR	30
C	50	F	College	Married	Nurse, Duty ward	Nurse, Duty ward	28
D	52	F	Master	Married	Nurse, OR	Nurse, OR	30
E	42	F	University	Married	Nurse, ER	NP, ER	21
F	46	F	University	Married	NP, ER	NP, ER	25
G	50	F	University	Married	Nurse, Duty ward	HN, ND	29
H	44	F	University	Single	Nurse, Duty ward	Nurse, Duty ward	25
I	53	F	Master	Married	HN, Duty ward	SUV, ND	34
J	54	F	Master	Married	HN, Duty ward	HN, DR	35

It can be seen from Table 1 that the participants were female nurses aged 42–55 (average age about 49 years old), with nursing experience of 21–35 years (average seniority about 24 years); nine were married and one was unmarried. During the SARS epidemic, they all worked in special units. Three participants were promoted to supervisor, but they still handled care-related business during the COVID-19 epidemic.

3.3. Data Collection

Snowball sampling combined with a semi-structured interview was applied in this study. The duration of data collection was between 2 February 2021 and 3 January 2022. However, in 2021, when the epidemic broke out in Taiwan, some participants were interviewed by communication software. The interview guideline was drawn up based on the purpose of the research and the literature review and included the following: (1) Can you talk about your experience when you faced the epidemic of SARS? (2) Regarding the COVID-19 epidemic, can you share your experience and feelings about it? (3) Can you talk about the differences between SARS and COVID-19? (4) Do you have any other things would like to share with me?

3.4. Data Analysis

In the data analysis, the narrations and extracted codes were discussed with other qualitative research experts in relation to the subject of study, and their indications were considered. An ethical consideration was obtaining permission from the ethics committee of the Guang Ten General Hospital Human Experiment Committee (IRB 10952), a written informed consent from the participants was obtained, and consideration was given to the participants' right to cancel their attendance in the research.

3.5. Research Rigidity

To ensure rigor and credibility, this study followed the standards proposed by Lincoln and Guba [17]. Data credibility was established through the selection of appropriate data collection methods (guidelines for semi-structured interviews) and the manner in which the researchers conducting the interviews were familiar with the evidence of SARS and COVID-19. Reliability was ensured by describing the data analysis in detail and providing direct references to reveal the basis on which the analysis was performed. The researchers coded the interviews independently of each other. Consistency of analysis was determined through a meeting to discuss initial findings, where emerging norms and themes were discussed until a consensus was reached. This study was maintained throughout the coding process. To improve the portability of the study results, a background description, participant selection, data collection, and the analysis process are provided.

4. Results

This study mainly explored the feelings and responses of nurses who had experience in caring for SARS patients in the face of COVID-19. In total, 10 participants were interviewed by researchers. The basic information of the participants is shown in Table 1.

According to the data analysis, the researchers summarized the research results into two major themes and four sub-themes. The two main themes were: facing the epidemic from the unknown to the known; and the experience from ignorance to proficiency. There were four sub-themes: the feeling of frustration and concern; bottlenecks and pressures in my work; my mission and support; and positive energy and camaraderie. The details are shown in Figure 1.

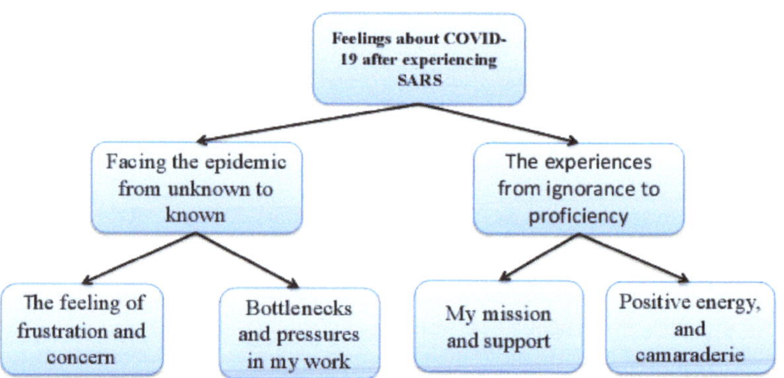

Figure 1. The themes and subthemes of study.

4.1. Facing the Epidemic from Unknown to Known

Facing COVID-19 is a way to face new challenges by stepping on past experiences. The feeling was different compared to the experiences of SARS. Participants shared their feelings about facing COVID-19 after experiencing the outbreak of SARS. The researchers classified the relevant information into the feeling of frustration and concern and bottlenecks and pressures in my work, the details of which are presented in the following section.

4.1.1. The Feeling of Frustration and Concern

In the face of the two epidemics, the participants felt worried and frustrated because they were not familiar with the disease; they also feared their mortality. Participants recalled the first time they faced the epidemic of SARS. The reason for their worry about SARS was fear:

> *At the time of SARS...it was an unknown disease...we really did not want to join the team to take care of patients, because we were also afraid...and a group of colleagues was crying over there ... scared because who knows what would happen... (Case H P2 L11)*

SARS was an unknown disease to the participants. Case H also described their feelings when facing SARS, particularly fear. In addition, the information received through the media was mostly negative, which also brought pressure on Case I in terms of care:

> *(SARS) I think the first thing I am not familiar with about this disease ... it is an unknown condition, I don't know how to control it ... and then I heard a lot of negative news from the media, they (media) said we did not have enough equipment ... they (media) a nurse pass away ... from the news the pressure is relatively high ... (Case I P3 L36)*

As a result of interview, Case I indicated her fear was not only the unknown disease but also the information from media. As mentioned previously, some researchers had indicated that media had mislead the public and medical teams and caused panic. Fortunately, the period of SARS was only a couple of months and everything was under control. However, 18 years after experiencing SARS, COVID-19 appeared without warning. Participants indicated how they felt when they heard about the COVID-19 epidemic:

> *... (SARS) Oh~ it's finally over, because I feel like ... oh my god! not again! ... When (COVID-19) really came later, the whole thing was (gasp)...how could it be like this~ but just bite the bullet...but this time it feels like it's nowhere in sight ... (Case H P5 L23)*

Case H decreased the intensity of their feelings when they faced COVID-19. The data of their interview showed that they had no choice and had less fear of the condition. Past experience allowed them to have a relatively low level of fear when facing the new epidemic. Case H pointed out with experience of SARS they had more protection for themselves when facing COVID-19. They said:

(COVID-19) the whole epidemic is when it is more serious...anyway, I had experience (SARS) of it, and I am not afraid...I just told myself that this time the protection should be better, and the way of infection is different. (Case H P6 L36)

With prior experience, Case H shared her experiences of facing COVID-19. However, the occurrence of the nosocomial infection of COVID-19 and the infection of nursing staff, the spread of the media, and the reprimands of the public happened. In particular, the shadow of the hospital closure during the SARS period still affected Case D:

... in fact, (SARS) Heping Fuyou Hospital was closed. ... because of this (hospital closed), a young doctor died ... Everyone was panic!...and (COVID -19) Butao (hospital)...happened again, people pointed the finger at the medical staff, and this is really hard to guard against~...It's quite frustrating! (Case D P2 L33)

Whether it was the Heping Fuyou hospital closure during SARS, or the COVID-19 incident, the participants have had to bear a lot of pressure. They worried that the hospitals they serve would experience a similar situation. At the same time, the attitude and direction of public opinion would also change because of the conditions of the epidemics. The feelings of frustration can be seen in the content of the participants' interviews. However, participants did not only feel frustrated by the condition of two epidemics and some misleading information; because of their duty, their families were also the participants' biggest concern.

Participants mentioned that their family members tended to suggest they should resign their jobs. The participants also worried about infecting their family, especially those who were married with children. Case J mentioned that family members had asked them to leave their job due to concerns about the high risk of caregiving:

... (SARS) at that time, my child was very young, and even the elder relative said, resign your job; it's too dangerous ... I'm also afraid that if I infect my family members, ... the risk of infection is relatively high, the nurses left the job because their family wanted them to do it... (Case J P4 L27)

The unclear situation of SARS made the family worry about the safety of the participants. Conversely, the safety of the children at home was also a factor for participants to be concerned about not going home. In addition, there were nosocomial infections in both outbreaks, which made the participants worried that they might infect their family. The condition caused a lot of psychological pressure. Case F said:

... (during SARS) We were sad because of the incident at the Heping Fuyou Hospital ... we were afraid that we would infect our family, my kids were very young at that time...the pressure from the family was very high because you will say oh~...you see it is so serious...or ... you might think of quit the job! (Case F P4 L10)

Because of SARS, the participants were concerned about cluster infection. However, when the COVID-19 epidemic broke out, participants learned that nursing staff and their families had been infected due to nosocomial infection. They felt more empathetic and feared that they would follow in the same footsteps:

... This time (COVID-19) I saw the nurses (infected)...and their family were infected, and I felt really pitiful. I can understand that feeling... (Case B P12 L13)

I did not go home because I did not want to cause my families infected ... you know ... during SARS, they (school) asked my child stay at home ... they thought my kid was an infectious factor ... early stage of COVID-19 ... you see from the news, almost the same ... fortunately, this time the news from media was more acceptable ... people changed their attitudes ... (Case I P10 L10)

As a first-line nurse, participants had their duty. However, they had different feeling about the two outbreaks. Because of the unknown and the media's misleading messages, participants felt panic during the time of SARS. In the face of COVID-19, due to different nursing experiences and disease attributes, the degree of fear was very personalized.

However, family and children were their main concerns. Their work caused their families to be ostracized by the public or by children in schools. The results showed participants' grievance and helplessness.

4.1.2. Bottlenecks and Pressures in My Work

Whether it is SARS or COVID-19, it is a high-risk job for first-line nurses, who need to provide dedicated and centralized care for infected patients. The two epidemics in different periods caused different pressures for the participants because of the different care recipients; additionally, the pressure on patient care in the different epidemics was also different. For example, Case E indicates:

> *... at that time (SARS),...working in the isolation room... work alone!...I was afraid of being infected...differently, (COVID-19)...we take care of patients with seriously condition, and the high infection rate...it is actually physical and mental stress ... I'm tired...SARS was just a few months, but this time we really fought for too long!* (Case E P5 L19)

Case E pointed out the differences of working with SARS and COVID-19. Working alone and a long period of fighting with epidemics caused them different pressures in their work. Apart from the duration of the epidemics, the participants also mentioned manpower and resources.

As previously mentioned, when the SARS outbreak occurred, there was no relevant information about the unknown disease. In particular, information on the disease and the policies were not very clear, so the participants were more likely to feel frustrated. The COVID-19 epidemic was based on the experience of SARS, and the command center set up by the government created all the information. The experience of SARS also made everything easier. Case A mentioned their experience of the two outbreaks. The first thing mentioned in Case A is the nursing manpower:

> *... (SARS) ... there was only one nurse on duty, there was no way to ask for help. Of course, inside the heart would become more anxious and panic...(COVID-19) ... everyone will work together as a team ... know how ... know why ... m! it's much better.* (Case A P20 L2)

Case A shared their experiences during the SARS and COVID-19 epidemics. They mentioned their working conditions in both epidemics. From their viewpoint, the manpower and cooperation with colleagues was important. Case D also expressed the importance of the team, which can support everyone through the situation together, and said:

> *... (SARS)When you were there, you would be very lonely...everyone had to support each other, otherwise it is really hard, and everyone was panicked!* (Case D P9 L3)

Teamwork was mentioned by many participants, and they thought it was an important factor to support the first-line nurses in continuing their work. In addition, it was indicated by some participants that relevant policies, procedures, and training, as well as on-the-job education, were particularly important and could alleviate the feeling of fear.

> *... (SARS) at that time, I didn't know how to use N95...to be honest, the equipment in our hospital was not enough, it was the first time, everyone has no experience! ... no SOP at that time ... and now (COVID-19)...education and training are quite frequent ... It's very important ... I think it will really reduce the staff's great pressure at work.* (Case A P3 L30)

The experience of SARS made everyone aware of their shortcomings, so the planning of relevant training improved the basic capabilities of medical staff and enhanced their sense of security when caring for patients. In addition, Case C also indicated that on-the-job training was carried out regularly after the SARS incident to "prevent fools".

> *... After the SARS incident,...infection control will teach you! And they will give you a standard of picture to teach you how to put on and take of the protective suit, and take a*

test how to put on and take off every six months . . . we call it the "preventing fools" . . .
(Case C P4 L5)

In addition to education and training, Case F also mentioned about the unknown information, lack of policies, and insufficient direction; they felt helpless because they were afraid of being infected. During the COVID-19 period, protective materials have improved significantly, and the protective equipment is sufficient.

. . . it was chaotic at the beginning of SARS, there was no system, and when there was nothing to follow, everyone would be afraid...This time COVID-19,...our first-line nurses have first-hand information such as SOP, policies . . . I think it was not as frustrating as SARS...the protective measures are relatively complete. (Case F P2 L31)

The participants mentioned the importance of the policies and the training program, as well as sufficient manpower and equipment. With the accumulation of experience and the improvement of infection control measures, during this epidemic period, the hospital's protective equipment was sufficient. With "know how" and "know why", it is clear that information given to the first-line nurses means they have less pressure at work.

4.2. The Experience from Ignorance to Proficiency

When caring for patients during SARS, they felt ignorant, helpless, and afraid at the beginning, but due to their relevant experience their coping methods matured when facing COVID-19. From the data of the interviews, the researchers divided this stage into "my mission and support" and "positive energy and camaraderie".

4.2.1. My Mission and Support

From the participants' point of view, sacrifice is the expectation and aura that nurses have always had. Many expectations have been placed on them. Since the outbreak of the SARS epidemic began in public hospitals, they could only accept instructions to take care of patients in the face of the epidemic. During COVID-19, the epidemic has been longer and more challenging compared to SARS, but, due to their relevant nursing experience, plus their passion for nursing, they are still willing to devote themselves to caring for patients:

(SARS) the patients are very disturbed, and they will frequently ask, "Miss, am I severity?...Is it like the one reported on TV? Will I die? "...then you need to appease the patient's feeling, and then you have to be strong enough . . . (Case A P3 L8)

During the interviews, Cases A and I also expressed that tourists carried the virus into Taiwan in the early stages of the spread of COVID-19. Faced with these patients, there was ambivalence, and they also figured out their problem of looking after people from overseas:

. . . (COVID-19) I just want to say "Why should I care about you people?...foreigners come to Taiwan and carried the virus..., I feel...just...but I told myself that he is sick and is a patient, and he should be treated fairly, I don't want to say that he is excluded. But do you know what's my problem?...my language . . . (Case A P4 L9)

. . . the foreigner...I don't know . . . this time (COVID-19) . . . majority of the incidences were of COVID-19 were carried by foreigner, oh, yes, I did have to look after them but what caused me headache was language (laughing) I don't think my English is good enough to communicate with them . . . (Case I P7 L19)

The participants described their commitment to the job. With their expectation and aura, they have to hide their fear and weakness from patients. They also have to hide their anger to face the tourist diagnosed with COVID-19. It is not easy and reflects the difficulty of those first-line nurses during epidemics. In addition, from the results of the interviews, the participants also presented the weakness of their language ability. This also caused the participants a feeling of pressure. However, in order to put themselves in a better condition to take care of their patients, they shared their experiences of how they adjusted themselves.

Stress...I'm trying to face it positively...The first thing you should do is to be healthy! That is, the diet and nutrition must be sufficient, and then there must be enough sleep. (Case G P6 L36)

Case G said that eating well, sleeping well, and protecting themselves, were their coping methods to survive the epidemic. The participants also indicated that they monitored their health and maintained a positive attitude when facing the epidemic, as well as self-protection and washing their hands, which helped them feel at ease.

. . . You have to learn how to protect yourself. If you feel that you have done a good job of protecting yourself, your mood will be more stable ~ feel at ease. (Case G P7 L4)

I...I will try to avoid eating with my colleagues,...I think I have taken protective measures well, but I am not necessarily know whether my colleagues have done well or not . . . (Case F P9 L11)

Nurses know that their work risks are high, so they will protect themselves, including taking protective measures and washing their hands properly. In addition, in order to protect family members, the nursing staff will choose to clean themselves before going home or reduce contact with family members, as Case A said:

(COVID-19) . . . We are afraid of infecting our family members, so before going home, we will take shower... (Case A P13 L9)

On the other hand, the support from family provides the nursing staff with great strength to help them through the epidemic period. The degree of support from family members determines whether the nursing staff are willing to continue to struggle in the workplace. They said:

. . . I once told my husband that I am taking care of this kind of case, ...I am afraid of infecting him, but my husband told me "Don't be afraid, if we are quarantined, we will be together" (Case A P14 L20)

. . . my husband is very supportive of me...he believes that I will do well... My mother look after my children...During SARS, my family was very supportive, and then...when faced with such a situation (COVID-19) again, their attitude was still the same. (Case F P8 L2)

In the face of the two epidemics, the participants indicated that good management of health, including exercise, nutrition, and sleep, would help to prevent them being infected. Support from their family is the most important way to support nursing staff through the epidemic. Although their families are still worried for their safety, they also respect their profession and give them great support.

4.2.2. Positive Energy and Camaraderie

As mentioned earlier, during the SARS period, due to the underdevelopment of communication software, many messages were unclear, and there was a feeling of isolation because of working alone. With the development of information nowadays, communication software and related information can be used to convey care or correct information, which can bring positivity.

. . . In the beginning (SARS) there were only mobile phones~ and then the phone...the feeling is lonely. ...nowadays technic is more developed~...it's more...not feeling of isolated... (Case D P3 L25)

(COVID-19) the power of the internet is quite strong... there will be social learning . . . there are some positive reports. ...everyone joins in the grand event...I think that it is much better than before. (Case I P12 L38)

As mentioned in Case I, the Internet is very powerful. The participants mentioned that they used the first-line group to convey care, and they also used the power of the Internet to spread a positive atmosphere. Compared with the SARS period, they have more positive energy to carry out their patient care and face the outbreak. When conveying messages using the Internet, the participants mentioned mutual help among colleagues.

> *When I saw my colleagues, I cried!...They said, "why you are crying? We are here for you."...you are not alone . . . .. I mean that the team can help each other~ . . .* (Case A P4 L25)

> *. . . Colleagues...caring for each other, in fact, I also think it is very important to encourage each other...every time I go to the ward, I will remind each other to be carefully . . .* (Case H P4 L6)

> *. . . I think our dean is very kind. He leads us~ no criticize . . . everyone is working very hard in epidemic prevention, and we must give encouragement to these infected people. In fact, the hospital's attitude towards us means that we should give care to these people and go through with them together.* (Case G P19 L3)

Nurses on the first line often have to face many things. As mentioned by the participants, the dean's approach of "no criticize" moved and warmed the participants. According to the results of the interviews, the participants shared their experiences and feelings when they faced the epidemics. It can be seen that support from the public and their families is very important to them. As well as the "no criticize" approach, these are key points that keep them going.

5. Discussion

From the results of the interviews, the participants indicated that, because of the unknown, people were afraid of SARS. For example, the participants said: *"Because it is an unknown disease, ...I am afraid because it is unknown"* and *"I feel panic! Not at all. I do know anything! I only know that it will kill people (SARS)."* This shows that when facing SARS people felt panic and fear because they did not know anything about the disease, only that there was a high mortality rate. As Maunder et al. [14] mentioned, the sources of stress for medical staff during the SARS period included uncertainty about the disease and increased mortality, which is consistent with the research results.

When faced with COVID-19, the feeling of frustration was reported by the participants. Due to previous nursing experience and the severity of the disease, the degree of and the reasons for the fear and stress were different. As the participants said: *"This COVID-19 is more terrifying than the previous one, because of the mutation"* and *"It may be because of SARS."* However, they were less likely to feel so frustrated and overwhelmed. This shows that due to the participants' experience of SARS, the degree of fear of COVID-19 is different; however, they still have the feeling of fear, worry, panic, and other negative feelings. The results of this study are consistent with the fact that many research studies have mentioned that there will be psychological distress during the epidemic [18,19].

In addition, family members are the most important factor affecting the participants. The participants mentioned that they are worried that they could be a source of infection to their family. For example, the participants said: *"If you have children, you will be more worried about whether you will infect the children..."* This reflects the research results that indicated the factor of worrying about infecting family members [5,9].

Moreover, during the SARS period, there were also people who wanted to leave their jobs because of their family members. For example, the participants said: *"The wave of resignation...many of them are because of parents' concerns...because they think the risks of death are too high"* and *"At that time, my children were still very young, ...the elder relative said that it's so dangerous, maybe I have to resign my job."* During the COVID-19 period, there have been few resignations, which is different from the SARS period. Wei [20] mentioned that during the SARS period the nurses were under pressure and family members were reluctant to leave their jobs, which led to a shortage of manpower. Conversely, in the period of COVID-19, protecting and caring for first-line nurses is the most important mission for the administration [19].

According to the results of the interviews, the married participants had a high percentage of negative feeling about the epidemic compared to those who were unmarried, especially those with children. For example, the participants said: *"At that time (SARS)...I was younger... There are not too many things to be afraid of...Now that I have children~...."* Sun et al. [21] mentioned that caregivers with children at home will be particularly worried, but there are also studies pointing out that married people have a more positive attitude toward COVID-19 [21], which shows that this feeling is related to personal experience.

Moreover, regarding the participants' experience during the SARS period, there were some factors that caused them to feel a kind of frustration, including working alone, insufficient protective materials, working with scant information, and unclear policies and lack of knowledge. Fortunately, those uncertain conditions have improved significantly during the COVID-19 period. For example, one participant said: *"(SARS) didn't know how to wear the N95 at that time? ...there was no standard...education and training...like now (COVID-19)...this time it has been effectively improved."* The participants' reports reflect some researchers' viewpoints. For example, Guo et al. [9] and Maunder et al. [14] pointed out that frequent policy changes and insufficient equipment are factors that cause first-line nurses physical and mental stress. In addition, Feng et al. [5] and Labrague et al. [22] stated that on-the-job education and accurate information about the disease can reduce first-line nurses' fear and negative emotions.

A sense of mission has been held in Taiwanese nursing society since the 1900s until now. In this regard, all participants carry the sense of mission to care for patients from different parts of the world during epidemics. Many participants mentioned: *"I still have the responsibilities and missions . . . as a nurse, and I don't want to say that they are excluded."* and *"I continue to work in nursing work ~ mission drives me forward!"* The issues that nurses face were also indicated by Guo et al. in 2005. Their [9] study stated that, during the epidemic of SARS, uniforms are an honor for nurses but they also give them the responsibility and pressure to perform the job and care for patients.

Transcultural nursing practice is another challenge for participants. With the change in society, Taiwan has become a multi-culture country. COVID-19 has given the participants the pressure to face language problems. They maintained that: *"different languages are really the most stressful for us at present."* In responding to participants' pressure, Chuang et al. [23] suggested that in response to the advent of the era of globalization, Taiwan's nursing profession should also step into internationalization. It has been suggested that the foreign language ability of nursing staff should be improved.

Moreover, the support from family has been most important for the first-line nurses during the epidemic. On the other hand, first-line nurses' safety is the main concern of their families. Their families also respect the participants' duty of profession and they provide great support. For example: *"my husband...he gave me a lot of psychological support"* and *"I think they support me a lot . . . from their hearts."* Family support is the key factor for nurses in fighting epidemics, which has also been mentioned by Sun et al. [21]. In 2005, a study by Guo et al. investigated the work-related stress and coping behaviors during the SARS outbreak period among emergency nurses and found that the participants' coping method was to feel and receive support from their family. The results of this study also reflect the viewpoints of Sun et al. [21] and Guo [9].

Apart from the support from family, mutual support among colleagues and positive encouragement from leaders and administrations create a positive atmosphere for participants: *"It's very supportive the situation like this . . . colleagues are here with you . . . , and you feel very touched . . . "*, *"The emergency director...is very positive, ...very supportive of us."* and *"The attitude by the hospital is that we want . . . our bosses be accompany with us. we are working together . . . ."* Pan et al. [10] mentioned the spirit of solidarity and the support from important others. Sun et al. [21] also mentioned that the hospital has a reward and welfare system to support and encourage nurses, and the encouragement of colleagues also makes nurses feel happy.

5.1. Limitations

The purpose of this study was to explore nurses' feelings who had experienced both SARS and COVID-19. Therefore, the number of participants was limited. In addition, the experience of SARS was 18 years ago, so there may be some memory bias in the results.

Furthermore, the duration of the data collection was between 2 February 2021 and 3 January 2022. These participants had not experienced the outbreak of the epidemic in May 2022. The participants refused a second interview because a limit to speech was announced by the hospital administration. The investigation of nurses' feelings during the May 2022 outbreak could be a further study for anyone who is interested.

5.2. Suggestions for the Future

1. Policy: a clear policy is very important for first line nurses. It also protects them to face to challenges of epidemics. It is seen that from SARS to COVID-19, the government agency in Taiwan had big improvements. However, in the early stage of COVID-19, the first line nurses and their families were still affected by the changeable conditions. As the result, the government agency has to think about the policies to support and protect the frontline soldier for the challenges in the unknown future.
2. Equipment: sufficient epidemic prevention equipment will affect the willingness of first-line nurses to care for patients as well as the safety of first-line nurses. From SARS to COVID-19, the government agency and administration in hospitals might have to consider to prepare enough materials to prevent future challenges.
3. Media: media are very important resources for publics especially during outbreaks. The media is a major source from which people perceive risk. As a result, the health authority should construct mechanisms or SOP to remind follow-up preventative of risk after these two evidences.
4. Physical and psychological health management for first-line nurses: it is recommended that medical and psychological professional consultants, such as doctors or psychologists, are available to support the first-line medical team.
5. On-the-job education: major institutions arrange infection control education and training regularly. It is recommended to internalize the education and training of wards and invite senior nursing staff with relevant nursing experience to serve as lecturers.
6. Strengthen the foreign language ability and information retention of nursing staff: as society is more globalized, a multicultural nursing knowledge is necessary.

6. Conclusions

Two outbreaks have caused serious harm in the medical system and the general public's health, especially with COVID-19 still unabated. Although the condition of an epidemic cannot be predicted, these two experiences can also be learned from, including the preparation of epidemic prevention materials and regular on-the-job education, especially relevant knowledge. Certainly, government decision-making capacity is also required.

In addition, the media is a useful resource, but it can also cause panic and stress among the population. The reaction of the public has a direct effect on nursing staff, and the reaction of the public is mostly based on the media. Therefore, proper media guidance is one of the best stress responses for nursing staff. The restriction and use of the media can test the wisdom of the competent authorities.

Finally, it is also necessary to consider supporting facilities related to the hospital, such as nosocomial infection prevention, sufficient manpower, and a rotation system, which can allow nursing staff to lower their levels of stress.

Author Contributions: Conceptualization, H.-L.L.; data curation, H.-L.L. and L.-C.L.; formal analysis, P.-J.C. and L.-C.L.; investigation, P.-J.C.; methodology, H.-L.L., P.-J.C. and L.-C.L.; project administration, L.-C.L.; supervision, L.-C.L.; validation, H.-L.L.; writing—original draft, H.-L.L. and P.-J.C.; writing—review and editing, H.-L.L. and L.-C.L. All authors have read and agreed to the published version of the manuscript.

Funding: This research received no external funding.

Institutional Review Board Statement: This study was approved by the Kuang Tien General Hospital Human Experiment Committee on 7 January 2021. The number is: Person 10952.

Informed Consent Statement: Informed consent was obtained from all participants involved in the study.

Data Availability Statement: The information presented in this study is available on request from the corresponding author.

Conflicts of Interest: The authors declare no conflict of interest.

References

1. Fang, Y.C. Experiences in SARS Nosocomial spreading. *J. Radiol. Sci. Introd.* **2004**, *29*, 191–195. Available online: https://www.airitilibrary.com/Publication/alDetailedMesh?DocID=10188940-200408-29-4-191-195-a&PublishTypeID=P001 (accessed on 15 December 2022).
2. Liao, H.C. The dilemma of being hospital workers: The SARS crisis at hospitals of Taiwan. *Formos. J. Med. Humanit.* **2008**, *9*, 51–58. [CrossRef]
3. Wu, Y.C. Standardized operation process for communication with mass media during disease outbreaks: Based on experiences from SARS. *Taiwan J. Public Health* **2007**, *26*, 241–249. [CrossRef]
4. The Ministry of Health and Welfare, and the Bureau of Disease Control. Severe Acute Respiratory Syndrome. 2018/12/1. Available online: https://www.cdc.gov.tw/Category/Page/Kou_i6ATU8jUnmKlAORhUA (accessed on 15 December 2022).
5. Feng, M.C.; Wu, H.C.; Lin, H.T.; Lei, L.; Chao, C.L.; Lu, C.M.; Yang, W.P. Exploring the stress, psychological distress, and stress-relief strategies of Taiwan nursing staffs facing the global outbreak of COVID-19. *J. Nurs.* **2020**, *67*, 64–74. [CrossRef]
6. Garfin, D.R.; Silver, R.C.; Holman, E.A. The novel coronavirus (COVID-2019) outbreak: Amplification of public health consequences by media exposure. *Health Psychol.* **2020**, *39*, 355–357. [CrossRef] [PubMed]
7. The Ministry of Health and Welfare, and the Bureau of Disease Control. COVID-19 Epidemic Prevention Zone and Latest Information. 202212/24. Available online: https://www.cdc.gov.tw/ (accessed on 25 December 2022).
8. Ke, Z.Y.; Xu, S.M.; Hong, M.N.; Duan, Y.C.; Lin, M.C.; Shi, C.K.; Zhang, Z.C.; Lu, M.L. Investigation of the first community cluster infection of the Delta variant of COVID-19 in Taiwan in 2021. *Taiwan Epidemiol. Bull.* **2022**, *38*, 75–80. [CrossRef]
9. Kuo, C.N.; Lee, B.O.; Lee, S.H. Work-related stress and coping behaviors during SARS outbreak period among emergency nurses in Taiwan. *Chang. Gung. Nurs.* **2005**, *16*, 139–151. [CrossRef]
10. Pan, S.M.; Feng, M.C.; Wu, M.H.; Chiu, T.Y.; Ko, N.Y. Voices from the frontline: Nurses' impact and coping during the 2003 SARS outbreak in southern Taiwan. *J. Evid. -Based Nurs.* **2005**, *1*, 149–156. [CrossRef]
11. Wang, S.I.; Yaung, C.L.; Liao, H.E.; Chiang, H.C. Media: A double-edged sword in the disease control. Content analysis of new report on SARS even in Taiwan. *Health.* **2016**, *6*, 1760–1766. [CrossRef]
12. Lee, S.; Juang, Y.; Su, Y.; Lee, H.; Lin, Y.; Chao, C. Facing SARS: Psychological impacts on SARS team nurses and psychiatric services in a Taiwan general hospital. *Gen. Hosp. Psychiatry* **2005**, *27*, 352–358. [CrossRef] [PubMed]
13. Huang, L.C.; Sun, J.J.; Hsiao, C.T.; Liu, T.Y. The knowledge, attitudes and stress of nurses in caring patients with Novel Coronavirus (COVID-2019): An exploratory descriptive study. *Taipei City Med. J.* **2021**, *18*, 445–459. [CrossRef]
14. Maunder, R.; Hunter, J.; Vincent, L.; Bennett, J.; Peladeau, N.; Leszcz, M.; Sadavoy, J.; Verhaeghe, L.M.; Steinberg, R.; Mazzulli, T. The immediate psychological and occupational impact of the 2003 SARS outbreak in a teaching hospital. *Can. Med. Assoc. J.* **2003**, *168*, 1245–1251. Available online: https://pubmed.ncbi.nlm.nih.gov/12743065/ (accessed on 25 December 2022).
15. Chen, C.P.; Huang, W.Y.; Wang, S.C.; Yeh, C.C.; Hsu, S.H.; Tang, P.H.; Cheng, S.H.; Cheng, C.Y.; Lin, Y.C. Effective control of a COVID-19 outbreak in a teaching hospital: A single-center experience. *J. Infect. Control.* **2022**, *32*, 231–245. [CrossRef]
16. Chen, Z. Conservational medicine: The only way to avoid emerging infectious diseases—The relationship between environmental changes and infectious disease outbreaks. *Harvest* **2020**, *70*, 8–14. [CrossRef]
17. Lincoln, Y.S.; Guba, E.G. *Naturalistic Inquiry*; Sage: Thousand Oaks, CA, USA, 1985.
18. Bai, Y.; Lin, C.C.; Lin, C.Y.; Chen, J.Y.; Chue, C.M.; Chou, P. Survey of stress reactions among health care workers involved with the SARS outbreak. *Psychiatr. Serv.* **2004**, *55*, 1055–1057. [CrossRef] [PubMed]
19. Chong, M.Y.; Wang, W.C.; Hsieh, W.C.; Lee, C.Y.; Chiu, N.M.; Yeh, W.C.; Chen, C.L. Psychological impact of severe acute respiratory syndrome on health workers in a tertiary hospital. *Br. J. Psychiatry* **2018**, *185*, 127–133. [CrossRef] [PubMed]
20. Wei, F.; Lee, Y.; Hsieh, H.; Huang, W.; Hu, H. Association of SARS experience with physical and mental health of healthcare organization employees in the early stages of COVID-19 pandemic. *Hospital* **2020**, *53*, 38–48. Available online: https://www-airitilibrary-com.proxyone.hk.edu.tw:8443/Publication/alDetailedMesh?DocID=P20190424001-202012-202101 070031-202101070031-38-48&PublishTypeID=P001 (accessed on 28 December 2022).
21. Sun, N.; Wei, L.; Shi, S.; Jiao, D.; Song, R.; Ma, L.; Wang, H.; Wang, C.; Wang, Z.; You, Y.; et al. A qualitative study on the psychological experience of caregivers of COVID-19 patients. *Am. J. Infect. Control.* **2020**, *48*, 592–598. [CrossRef] [PubMed]

22. Labrague, L.J.; de Los Santos, J. Fear of COVID-19, psychological distress, work satisfaction and turnover intention among frontline nurses. *J. Nurs. Manag.* **2020**, *29*, 395–403. [CrossRef] [PubMed]
23. Chuang, H.L.; Wang, C.C.; Kuo, P.C. The Internationalization of the nursing profession in Taiwan. *J. Nurs.* **2011**, *58*, 99–104. [CrossRef]

International Journal of
*Environmental Research
and Public Health*

MDPI

Article

Transformational Leadership and Emotional Labor: The Mediation Effects of Psychological Empowerment

Pengfei Cheng *[iD]**, Zhuangzi Liu and Linfei Zhou**

School of Economics and Management, Xi'an University of Technology, Xi'an 710054, China
* Correspondence: pfcheng@xaut.edu.cn

Abstract: In order to survive the fiercer competition, more and more service firms emphasize front-line employees' role of creating excellent customer experience by displaying positive emotions during the service interactions. However, the underlying mechanisms for the relationship between transformational leadership and front-line employees' emotional labor remain unclear. Drawing upon the conservation of resources (COR) theory, this study develops a conceptual model in which transformational leadership influences front-line employees' emotional labor through the mediator of psychological empowerment. By collecting data from 436 employees in five call centers, we tested our model and hypotheses through PROCESS 3.3 macro for SPSS developed by Hayes. The results show that transformational leadership shows positive and negative effects on deep acting and surface acting, respectively. The positive effect on deep acting is partially mediated by psychological empowerment, while the negative effect on surface acting is fully mediated by psychological empowerment. Specifically, two dimensions of psychological empowerment (impact, self-efficacy) play negative mediating roles between transformational leadership and surface acting, while impact, self-determination, and self-efficacy play positive mediating roles of transformational leadership and deep acting. The findings advance our understanding about how transformational leadership influences front-line employees' emotional labor by introducing psychological empowerment as a mediator.

Keywords: emotional labor; transformational leadership; psychological empowerment

check for
updates

Citation: Cheng, P.; Liu, Z.; Zhou, L. Transformational Leadership and Emotional Labor: The Mediation Effects of Psychological Empowerment. *Int. J. Environ. Res. Public Health* **2023**, *20*, 1030. https://doi.org/10.3390/ijerph20021030

Academic Editor: Masahito Fushimi

Received: 15 December 2022
Revised: 2 January 2023
Accepted: 3 January 2023
Published: 6 January 2023

1. Introduction

As the fast rise of the service economy in China and the competition becomes fiercer, more and more service firms are coming to realize the pivotal role of front-line employees in creating an excellent service experience. Front-line employees are required to display appropriate emotions during their interactions with customers [1]. In her seminal work, Hochschild [2] defined emotional labor as front-line employees' expression of expected emotions during service encounters. Employees can take two different strategies to perform emotional labor: surface acting and deep acting. Surface acting involves simulating emotions that are not actually felt. In contrast, deep acting involves attempts to actually experience the emotions one is required to display. Considerable evidence indicates that deep acting is more likely to lead to positive outcomes than surface acting [3–6].

Given the interactive nature of service process, how front-line employees display their emotions during service encounters determines the service performance. A vast body of research has given interest to identifying the antecedents of the emotional labor. One of the research streams on the antecedents of emotional labor focuses on customer behaviors, such as customer incivility [7] and customer participation [8]. However, another stream of research focuses on front-line employees' individual differences, such as dispositional traits [9], self-verification striving [10], and positive affectivity [11]. Due to the tight connection between leaders and front-line employees, a growing number of studies recently have shown interest in exploring the effects of leadership on emotional labor [12,13].

Transformational leadership is defined as the ability of managers to provide followers with challenging goals, motivating them to perform beyond the specified expectations [14]. Prior research suggested that transformational leadership has important effects on emotional labor [13,15]. However, the mechanism between transformational leadership and emotional labor has received little attention in previous literature.

To understand how different emotional labor strategies of front-line employees are influenced by transformational leaders, we examined the mediation effect psychological empowerment on the relationship between transformational leadership and emotional labor. Psychological empowerment, which involves employees' active orientation to their work role, is an important determinant of employee behaviors [16]. Through being empowered, front-line employees could get more internal resources in their job [17]. Previous research has shown that psychological empowerment could mediate the effects of the transformational leadership on employees' work outcomes [18–21]. However, to our best knowledge, few empirical studies have examined the potential mediating mechanisms of psychological empowerment through which transformational leadership influences front-line employees' emotional labor. Therefore, to address this gap, the purpose of this study is to unveil the underlying processes responsible for the effects of transformational leadership on front-line employees' emotional labor by introducing psychological empowerment as mediator.

To summarize, we proposed a mediation model in which transformational leadership influences front-line employees' emotional labor by improving psychological empowerment. Different from previous studies which mainly consider psychological empowerment as a unidimensional construct [22–24], the current study tries to further explore how transformational leadership influences emotional labor through four dimensions of psychological empowerment differently: meaning, impact, self-determination, and self-efficacy. By doing so, we may further explain the linkage between transformational leadership and employee's emotional labor and shed light on how managers might effectively interfere with front-line employees' emotional labor. In the following sections, we explicate the theorical background of transformational leadership and emotional labor and consider building bridges from four dimensions of psychological empowerment to the relationship between the prior two. We then test our hypotheses using a sample of 436 employees. Finally, we report the results and discuss the implications for theory and practice, offering constructive and specific inspiration for the service industry.

2. Theoretical Model and Hypotheses

2.1. Emotional Labor of Front-Line Employees

Emotional labor refers to front-line employees obeying their organizational expectation by regulating and displaying appropriate emotions during service encounters [25]. According to Hochschild [2], employees perform emotional labor by taking two different strategies: surface acting and deep acting. In surface acting, front-line employees comply with display rules by suppressing their inner feelings and expressing feigned rather than genuine emotions. Conversely, in deep acting, front-line employees try to actually experience the emotions they are required to display [26]. According to Ashforth and Humphrey [27], employees devote greater psychic efforts to perform deep acting than to perform surface acting. Deep acting requires front-line employees to actively invoke their thoughts, images, and memories to alter how they feel. Many previous studies have indicated that these two different emotional labor strategies lead to different employee well-being and service performance [28–30]. Generally speaking, surface acting results in negative employee and organization outcomes, including burnout [31], low job satisfaction [32], and more stress [33]. In contrast, the consequences of deep acting are more positive. Deep acting can reduce employee emotional exhaustion and enhance emotional performance [34,35]. Hence, just as Pugh et al. [36] calls for enhancing the practice of 'good' emotional labor in employees by promoting the use of deep acting and discouraging the use of surface acting, researchers are increasingly shifting their attention to explorer the antecedents of emotional labor. In order to help service firms effectively manage their front-line employees' emotional labor, many researchers have identified antecedents of emotional labor from an organizational perspective, such as organizational fairness [37] and leadership style [15]. Noticeably missing from research attention is the mechanism by which transformational leadership influences emotional labor, despite Chi et al. [38] having indicated that transformational leadership is positively related to service employees' deep acting.

2.2. Transformational Leadership

Transformational leadership refers to a set of behavior which elevate the capacity of followers through transforming their values, beliefs, and attitudes [39]. Previous studies have identified four dimensions of transformational leadership: charisma, inspirational motivation, intellectual stimulation, and individualized consideration [14]. Specifically, charisma refers to the ability of leaders to generate trust, admiration, and emulation of followers [40]. Inspirational motivation is defined as an encouragement behavior of leaders to motivate their followers to understand the vision of organization and exceed the established performance standards [14,40]. Intellectual stimulation is a behavior that leaders provoke followers to challenge existing assumptions and to face the problems with more innovative and creative in their work [18,41]. Individualized consideration involves treating each follower as a unique one, providing them with learning opportunities, and offering personal attention on their needs and concerns [42]. Previous studies have indicated that transformational leadership might lead to positive employee and organizational outcomes, such as high job satisfaction [43], organizational citizenship behavior [44], and organizational culture [45–47]. From the perspective of employees, Saira [18] suggested that transformational leaders can improve employees' organizational citizenship behavior and reduce their turnover intention. Her research further illustrates that transformational leadership's effects on employee behavior (OCB) are mediated by psychological empowerment. Similar findings was provided by Stanescu [21] that empowerment acts is necessary and effective processes for transformational leaders to foster innovation behaviors among followers. These findings imply that psychological empowerment plays an important mediating role in transformational leadership's effects on employees' behavior. Thus, it is reasonable to infer that transformational leadership influences front-line employees' emotional labor through psychological empowerment.

2.3. Transformational Leadership's Effects on Psychological Empowerment

Transformational leaders encourage subordinate self- development, provide a vision for the future, and pay attention to the subordinate's needs by exhibiting four kinds of behaviors: idealized influence, intellectual stimulation, inspirational motivation, and individualized consideration [19,48,49]. This people-orientated leadership style will foster the development of employees' positive psychological and organizational behavior [50]. Since transformational leaders rely more on innovative manners to deal with problems of their subordinates, the positive attitude of employees toward leaders and organizations are fostered, such as trust [51]. Thus, transformational leadership contributes front-line employees to generating intrinsic motivation and resources in some form, such as psychological empowerment.

According to the transformational leadership theory, empowerment has been recognized as a positive strategic management practice. Transformational leaders make it possible for employees to participate in the process of decision-making, cooperation, and idea generation, which could make them feel more empowered in their work [21]. Thus, subordinates of transformational leaders believe in the positive impact they could make to the organization [18]. In addition, research has pointed that psychological empowerment is vulnerable to leadership [20,24,52,53]. The research of Aydogmus [16] shows that employees will feel more psychologically empowered when they perceive their leaders as transformational. In other words, these behaviors from transformational leadership will lead to a certain external stimulus, and this kind of environmental stimulation will form a perception of psychological empowerment through the internal evaluation of employees [54]. Therefore, it is inferred that transformational leadership—as an external stimulus that pays attention to employee development—will affect psychological empowerment of employees.

Psychological empowerment refers to the extent of individual's perception of empowerment [55]. Previous studies have shown that employees' perceptions of psychological empowerment consist of four dimensions: meaning, impact, self-determination, and self-efficacy [56]. Specifically, meaning suggests that employees believe in the significance of their work and the alignment between their work goals and their personal goals [55]. Impact is a degree that individuals influence organizational outcomes, such as strategies, operations, and management [55]. Self-determination reflects a control of decision-making, emotions and behaviors, or the extent of freedom employees have in their work [21,57]. Self-efficacy refers to one's confidence of his/her capacity to perform a job with competence [55,56]. First, inspirational motivation of transformational leaders motivates employees to transcend the current standard of organization, which makes them view more value and meaning in their job [51]. Meanwhile, individualized consideration by transformational leaders will heighten personal development of employees [58]. According to the conservation of resources (COR) theory, these conditions increase job resource of employees, which are related to the personal growth of them [59], leading more positive attitudes towards organization, and then trigger the sense of meaning for them. Second, in the preceding passages, transformational leaders actively encourage employees to believe their ability [18]. On the one hand, these attempts create true feelings for employees about their contributions to an organization [60]. This feeling promotes them to have beliefs in their capability to have a great influence on the organization. On the other hand, encouragement from transformational leaders increases the confidence levels of employees, resulting in their self-efficacy [61]. Finally, intellectual stimulation of transformational leaders encourages their employees to challenge the status quo, to be innovative, and to take more responsibility in their work [53]. Consequently, employees will feel more empowered to deal with problems with a high level of freedom [20,51], enhancing their self-determination. Thus, we propose:

H1a. *Transformational leadership is positively related to meaning.*

H1b. *Transformational leadership is positively related to impact.*

H1c. *Transformational leadership is positively related to self-determination.*

H1d. *Transformational leadership is positively related to self-efficacy.*

2.4. Psychological Empowerment's Effects on Emotional Labor

2.4.1. Meaning's Effects on Emotional Labor

According Grandey et al. [62], front-line employees' emotional labor, as a kind of discretionary behavior, is mainly derived by employees' intrinsic motivations. The meaning of the job, which serves as an important promotor of one's intrinsic motivation, can motivate front-line employees to fulfill their job duties [63]. For example, when front-line employees can derive meaning from their daily work, they are more likely to internalize their organizational emotion display rules. Hence, we can infer that front-line employees who can get meaning from their jobs are prone to take deep acting rather than surface acting to comply with display rules when interacting with customers. Furthermore, the meaning of a job can bring front-line employees positive emotions [64]. With a more positive emotional state, it will take front-line employees much less effort to adjust their real feelings. According to the conservation of resources (COR) theory, positive emotions, as a kind of personal resource, can guarantee front-line employees to take deep acting, which creates more personal resources [65]. In addition, front-line employees with a high level of intrinsic motivation have a high level of job competence, such as cognitive flexibility and emotional regulation abilities [66]. These kinds of abilities makes front-line employees easer to induce themselves to feel the expected emotions, namely performing deep acting.

Conversely, organizational emotional display rules could be a burden for front-line employees who cannot get meaning from their job, bringing them more cognitive limitations. For these employees, they must always pay attention to their internal and external contradictory emotional states and spend energy restraining negative emotions. On the other hand, weakness of intrinsic motivation may also make front-line employees unaware of how to actively adjust their cognitions when facing unreasonable treatment from customers. Under this condition, front-line employees are prone to take surface acting to deal with the pressure from customers. Thus, we propose:

H2a. *Meaning is positively related to deep acting.*

H2b. *Meaning is negatively related to surface acting.*

2.4.2. Impact's Effects on Emotional Labor

In the workplace, how employees perceive the desired effects of their behaviors and the extent to which it affects their organization is defined as work impact [67]. Employees who are aware of the impact of their work have stronger feelings of responsibility towards the organization. They pay more attention to their emotions expressed during the process of interaction with customers. Such responsibility makes front-line employees actively internalize organizational service norms and values. Especially in tough times, this can help front-line employees overcome negative emotions and perform deep acting. In contrast, front-line employees who perceive low level of their work impact usually have a weak feeling of responsibility towards their organization. As a result, during their daily work, they are not prone to genuinely alter their inner feelings and pretend to feel desired emotions, namely surface acting. Thus, we propose:

H3a. *Impact is positively related to deep acting.*

H3b. *Impact is negatively related to surface acting.*

2.4.3. Self-Determination's Effects on Emotional Labor

Self-determination means a selective cognition of individuals in initiating and regulating their own behavior [68], reflecting front-line employees' level of autonomy in the workplace [69]. Previous researches have indicated that employees with high self-determination are of higher levels of concentration, initiative, and resilience [55]. Under this condition, front-line employees will devote much more effort to modify their inner feelings so as to display genuine, organization-desired emotions. Additionally, for the front-line employees with high autonomy, they will feedback their organizations with high identification, attachment, and loyalty [70], which will lead front-line employees to take more positive actions, such as deep acting to cope with job problems [71].

In contrast, it is difficult for employees with low self-determination to get a high sense of control over their job. Under the condition of low sense of control, front-line employees perceive high uncertainty and take passive manners to cope with job roles, such as surface acting [6]. Furthermore, a low level of self-determination involves feeling helpless and a lack of psychological resources. According to the conservation of resources (COR) theory, front-line employees with low level personal resources are more likely to take surface acting, which requires less personal resources to cope with their job roles [19]. Thus, we propose:

H4a. *Self-determination is positively related to deep acting.*

H4b. *Self-determination is negatively related to surface acting.*

2.4.4. Self-Efficacy's Effects on Emotional Labor

Self-efficacy refers to the extent of belief or confidence that one can perform work activities successfully [72,73]. As a form of psychological resources, self-efficacy causes positive psychological states, including felt responsibility, which motivates front-line employees to invest more efforts to perform their job roles [74]. According to the conservation of resources (COR) theory, front-line employees working in a resourceful environment are more likely to take deep acting, which duplets more resources than surface acting. Additionally, previous research has indicated that employees with high self-efficacy feel more confident to cope with customer demanding and make a positive appraisal of their work environment [70]. In other words, self-efficacy causes front-line employees positive psychological states, such as experiencing positive emotions. Under this condition, it is much easier for front-line employees to regulate their inner feelings (deep acting) to comply with the display rules than the condition of experiencing negative emotions [75].

Conversely, for front-line employees with low self-efficacy, they will underestimate their capabilities, which makes them experience negative emotions, such as stress and anxiety [76]. Negative emotions, which violate the basic service norm, make front-line employees have to suppress their inner feelings and pretend to experience positive emotions, namely surface acting. Furthermore, self-efficacy, as a foundation of intrinsic motivation, drives front-line employees to work hard and expect desired performance. However, for front-line employees with low self-efficacy, they are prone to take less effortful faking process without altering how they feel, namely surface acting. See Figure 1 for a model overview. Hence, we propose:

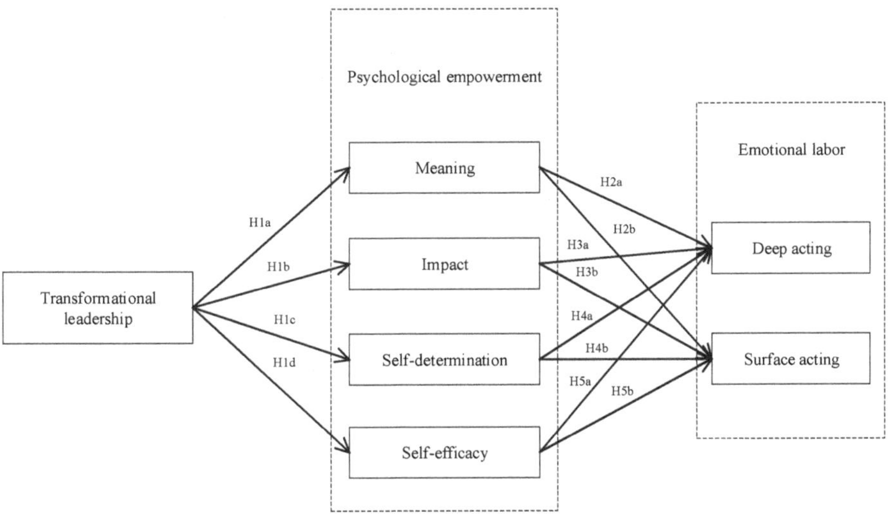

Figure 1. The conceptual model.

H5a. *Self-efficacy is positively related to deep acting.*

H5b. *Self-efficacy is negatively related to surface acting.*

3. Methodology

3.1. Sampling and Data Collection

In order to test the hypotheses, we carried out a survey to collect data from front-line employees of call centers. The call center was selected as our study context for its character of intense employee–customer interaction, which requires front-line employees to display positive emotions during the service encounter [77]. After selecting five call centers located in northwestern China, a total of 500 questionnaires were distributed on site to front-line employees with 100 questionnaires for each call center. The questionnaire was totally anonymous and took participants about 15 min to complete. When participants finished the survey, they were required to seal the questionnaire in an envelope by themselves and give it to the research assistants on site. Finally, 483 were returned. After deleting 47 incomplete ones, we obtained 436 valid questionnaires, indicating a response rate of 87.2%. The sample consisted of 436 employees (man = 30, woman = 406). The number of employees between 25 and 30 years was the largest (23.6%). Overall, 78% of employees had received a junior college education or above. In terms of tenure, the largest number of groups is one to three years (36.7%).

3.2. Measure Operationalization

For constructs involved in this study, we use previously established scales to measure. Because the original scales were developed in English, following the translation and back-translation procedures recommended by Brislin [77], we translated these measurement scales into Chinese. All measures except for demographic variables were reported on a five-point Likert scale, ranging from 1 (completely disagree) to 5 (completely agree).

Transformational Leadership. Transformational leadership was measured by a 20-item scale—the Multifactor Leadership Questionnaire (MLQ)—adapted from Bass and Avolio's [48]. MLQ consists of four correlated dimensions: charisma (8-item), inspirational motivation (4-item), intellectual stimulation (4-item), and individualized consideration (4-item).

Psychological Empowerment. Adapting from Spreitzer [56], psychological empowerment was measured via 12 items tapping four dimensions: meaning, impact, self-determination, and self-efficacy. Each dimension includes three items.

Emotional labor. Front-line employees' emotional labor was measured by using two 3-item scales, adapted from Brotheridge and Lee [78]. Both deep acting and surface acting were measured by three items.

Control Variables. To exclude potential confounding effects of front-employee demographics on emotional labor, we controlled for gender (0 = female; 1 = male), age (1 = under 21; 2 = 21 to 25; 3 = 25 to 30; 4 = 30 to 35; 5 = above 35), education (1 = Junior high school and below; 2 = High school; 3 = Undergraduate; 4 = Postgraduate and above), and tenure (1 = less than 1 year; 2 = 1 to 3 years; 3 = 3 to 5 years; 4 = 5 years and above).

3.3. Descriptive Statistics and Intercorrelations

The means, standard deviations, reliabilities, AVE, and correlations among the research variables are presented in Table 1. The Alpha coefficients for all constructs ranging from 0.862 to 0.932 indicate that the reliability is acceptable. Additionally, it can be seen that the significant correlations in the matrix are between the variables most proximal to each other in the hypothesized model.

Table 1. Mean, standard deviations, correlations, Cronbach alpha coefficient, and AVE.

	M	SD	α	1	2	3	4	5	6	7
1.Transformational leadership	3.668	0.747	0.932	0.642						
2.Meaning	3.848	0.765	0.908	0.431 ***	0.879					
3.Impact	3.250	0.893	0.872	0.461 ***	0.273 ***	0.835				
4.Self-determination	3.658	0.853	0.862	0.490 ***	0.393 ***	0.574 ***	0.823			
5.Self-efficacy	3.663	0.986	0.894	0.425 ***	0.324 ***	0.277 ***	0.309 ***	0.862		
6.Surface acting	3.654	0.892	0.900	−0.280 ***	−0.204 ***	−0.275 ***	−0.149	−0.343 ***	0.869	
7.Deep acting	3.704	0.935	0.898	0.474 ***	0.223 ***	0.391 ***	0.456 ***	0.311 ***	−0.035	0.867

Note: The square roots of AVE are presented in diagonal elements (bold values). n = 436; *: $p < 0.05$; **: $p < 0.01$; ***: $p < 0.001$.

3.4. Confirmatory Factor Analysis and Common Method Bias Testing

To examine the discriminant validity of the seven latent constructs, we first conducted a series of confirmatory factor analysis (CFA) with Amos 26.0 by utilizing maximum-likelihood estimation to test the measurement model (see Table 2). A total of four measurement models were compared by several fit indices: seven factors, three factors, two factors, and one factor models. The results suggest that the seven factors model fits the data best (χ^2 / df = 2.695, root mean square of approximation [RMSEA] = 0.062, comparative fit index [CFI] = 0.904, Tucker–Lewis index [TFI] = 0.894). Thus, the discriminant validity of the seven latent constructs is acceptable. Furthermore, the average variance extracted (AVE) of all constructs were between 0.642 and 0.879 (see Table 1), which exceeded the cut-off value of 0.5, indicating that the convergent validity of scales was acceptable.

Table 2. The results of confirmatory factor analysis.

Model	χ^2	df	χ^2/df	RMSEA	CFI	TLI
7 factors: TL, Mean, IMP, SD, SE, SA, DA	1705.791	633	2.695	0.062	0.904	0.894
3 factors: TL, Mean + IMP + SD + SE, SA + DA	4499.657	651	6.912	0.117	0.656	0.629
2 factors: TL + Mean + IMP + SD + SE, SA + DA	5176.653	653	7.927	0.126	0.596	0.565
1 factor: TL + Mean + IMP + SD + SE + SA + DA	5643.985	654	8.630	0.133	0.554	0.521

Note: n = 436; TL = transformational leadership; Mean = meaning; IMP = impact; SD = self-determination; SE = self-efficacy; SA = surface acting; DA = deep acting.

Since we used self-report methods to collect data, there may be common method bias, which may result in spurious relationships among the variables [79]. The results of Harman's single factor test have demonstrated that the first factor only accounts for 32.070% of the total variance, which indicated that common method bias was not present.

4. Results

To test the mediation effects of psychological empowerment on the relationships between transformational leadership and emotional labor, this study ran a series of regressions using the SPSS PROCESS macro developed by Hayes [80]. Specifically, following the suggestion of Hayes (2018), this study performed bootstrapping analysis with 5000 replications, which can generate 95% confidence intervals (CI) for total effects, direct effects, and indirect effects. The significance of mediating effect depends on whether CI contains zero or not.

First, this study tested the effects of transformational leadership (TL) on deep acting (DA) via the four dimensions of psychological empowerment. The results of mediation tests are summarized in Table 3. Coefficients of M-1 indicate that transformational leadership is positively related to deep acting (β = 0.465, p < 0.001). As mentioned earlier, we have hypothesized that transformational leadership influences deep acting through the meaning (Mean), impact (IMP), self-determination (SD), and self-efficacy (SE). Therefore, we examined the effects of transformational leadership on meaning, impact, self-determination, and self-efficacy, in M-2 to M-5. As expected, transformational leadership was found to be positively associated with meaning (β = 0.431, p < 0.001), impact (β = 0.406, p < 0.001), self-determination (β = 0.462, p < 0.001), and self-efficacy (β = 0.417, p < 0.001), supporting H1a to H1d. These findings demonstrate that transformational leaders in service firms could boost front-line employees' experience of psychological empowerment, which is consistent with the findings of prior studies.

Then, the results of M-6, which regressed deep acting on transformational leadership and four mediators (meaning, impact, self-determination, and self-efficacy) simultaneously, show that transformational leadership's positive effect (β = 0.304, p < 0.001) on deep acting remains. Three dimensions of psychological empowerment (impact (β = 0.106, p < 0.05), self-determination (β = 0.191, p < 0.001), and self-efficacy (β = 0.111, p < 0.05) are positively related to deep acting, supporting H3a, H4a, and H5a. However, meaning's effect on deep acting is not significant, rejecting H2a. The results of bootstrap analyses, summarized in Table 4, suggest that the CIs for impact (CI = 0.005, 0.116), self-determination (CI = 0.040, 0.198), and self-efficacy (CI = 0.006, 0.123) do not include zero and CI for meaning (CI = −0.075, 0.038) include zero. Taken together, the effect of transformational leadership on deep acting is partially mediated by impact, self-determination, and self-efficacy.

Table 3. The results of mediation tests.

Independent Variable		M-1 DA	M-2 Mean	M-3 IMP	M-4 SD	M-5 SE	M-6 DA	M-7 SA	M-8 SA
Control Variables	Gender	−0.105 *	0.070	−0.053	−0.061	0.058	−0.092 *	−0.091	−0.078
	Age	−0.076	0.082	−0.001	0.014	0.082	−0.085	0.026	0.048
	Education	−0.031	−0.069	0.049	−0.033	−0.064	−0.025	0.008	0.000
	Tenure	−0.068	0.064	−0.050	−0.036	0.040	−0.058	−0.061	-0.053
TL		0.465 ***	0.431 ***	0.406 ***	0.462 ***	0.417 ***	0.304 ***	−0.250 ***	−0.094
Mean							−0.032		−0.059
IMP							0.106 *		−0.172 **
SD							0.191 ***		0.079
SE							0.111 *		−0.232 ***
R^2		0.250 ***	0.198 ***	0.174 ***	0.213 ***	0.180 ***	0.313 ***	0.0767 ***	0.152 ***

Notes: TL = transformational leadership; Mean = Meaning; IMP = Impact; SD = self-determination; SE = self-efficacy; DA = deep acting; SA = surface acting; * $p < 0.05$; ** $p < 0.01$; *** $p < 0.001$.

Table 4. Total, direct, and indirect effects of TL on deep acting.

		Effect	Boot SE	Boot LLCI	Boot ULCI
Total effects		0.582	0.053	0.477	0.686
Direct effects		0.380	0.064	0.254	0.506
Indirect effects	TOTAL	0.202	0.060	0.088	0.327
	TL-Mean-DA	−0.017	0.029	−0.075	0.038
	TL-IMP-DA	0.054	0.031	0.005	0.116
	TL-SD-DA	0.110	0.040	0.040	0.198
	TL-SE-DA	0.055	0.033	0.006	0.123

Notes: TL = transformational leadership; Mean = Meaning; IMP = Impact; SD = self-determination; SE = self-efficacy; DA = deep acting.

Secondly, we tested the effects of transformational leadership (TL) on surface acting (SA) via the four dimensions of psychological empowerment. As presented in Table 3, coefficients of M-7 indicate that transformational leadership shows a negative effect on surface acting ($\beta = -0.250$, $p < 0.001$). Then, coefficients of M-8 suggested that after entering four mediators (meaning, impact, self-determination, and self-efficacy), the effect of transformational leadership on surface acting becomes non-significant ($\beta = -0.094$, $p > 0.05$). In addition, impact ($\beta = -0.172$, $p < 0.01$) and self-efficacy ($\beta = -0.232$, $p < 0.001$) are negatively related to surface acting, supporting H3b and H5b. However, neither meaning ($\beta = -0.059$, $p > 0.05$) or self-determination's effects ($\beta = 0.079$, $p > 0.05$) on surface acting are significant, rejecting H2b and H4b. In Table 5, the results of bootstrap analyses indicate that CIs for impact (CI =−0.158, −0.21) and self-efficacy (CI = −0.190, −0.055) do not include zero, while CIs for meaning (CI = −0.089, 0.030) and self-determination (CI = −0.029, 0.124) include zero. In addition, the direct effects of transformational leadership were not significant (CI = −0.246, 0.021). Taken together, the effect of transformational leadership on surface acting is fully mediated by impact and self-efficacy.

To summarize, for the effects of transformational leadership on emotional labor, all of our hypotheses are supported except H2a, H2b, and H4b (see Table 6). The results indicate that transformational leadership influences deep acting and surface acting through distinct mediating paths. Specifically, transformational leadership's positive effects on front-line employees' deep acting are partially mediated by impact, self-determination, and self-efficacy. In contrast, transformational leadership's negative effects on front-line employees' surface acting are fully mediated by impact and self-efficacy.

Table 5. Total, direct, and indirect effects of TL on surface acting.

		Effect	Boot SE	Boot LLCI	Boot ULCI
Total effects		−0.298	0.056	−0.409	−0.188
Direct effects		−0.113	0.068	−0.246	0.021
	TOTAL	−0.186	0.061	−0.311	−0.072
Indirect effects	TL-Mean-SA	−0.030	0.030	−0.089	0.030
	TL-IMP-SA	−0.083	0.035	−0.158	−0.021
	TL-SD-SA	0.043	0.038	−0.029	0.124
	TL-SE-SA	−0.116	0.035	−0.190	−0.055

Notes: TL = transformational leadership; Mean = meaning; IMP = impact; SD = self-determination; SE = self-efficacy; SA = surface acting.

Table 6. Hypothesis test results.

Number	Hypothesis	Result
H1a	Transformational leadership is positively related to meaning.	support
H1b	Transformational leadership is positively related to impact.	support
H1c	Transformational leadership is positively related to self-determination.	support
H1d	Transformational leadership is positively related to self-efficacy.	support
H2a	Meaning is positively related to deep acting.	reject
H2b	Meaning is negatively related to surface acting.	reject
H3a	Impact is positively related deep acting.	support
H3b	Impact is negatively related to surface acting.	support
H4a	Self-determination is positively related deep acting.	support
H4b	Self-determination is negatively related to surface acting.	reject
H5a	Self-efficacy is positively related deep acting.	support
H5b	Self-efficacy is negatively related to surface acting.	support

5. Discussion

The current study proposed and tested a conceptual model to explore the mechanism by which transformational leadership influences the emotional labor strategies of front-line employees. Our results demonstrate that transformational leadership exerts positive and negative effects on deep acting and surface acting, respectively. Psychological empowerment serves as a mediator of the linkage between transformational leadership and two emotional labor strategies. Specifically, the positive effect of transformational leadership on deep acting is partially mediated by three dimensions of psychological empowerment: impact, self-determination, and self-efficacy. The negative effect of transformational leadership on surface acting is fully mediated by two dimensions of psychological empowerment: impact and self-efficacy.

5.1. Contributions to Theory

Firstly, the results of this study help us better understand the psychological process by which organizational factors (e.g., transformational leadership) drive front-line employees' emotional labor. Although previous studies have paid attention to the effects of transformational leadership on front-line employees' emotional labor, such as Luo et al. [13], the psychological mechanism of these effects remain unclear. Grandey et al. [62] mentioned in their work that front-line employees' emotional labor is a form of discretionary behavior, which is driven more by intrinsic motivation. It is therefore reasonable to explain how transformational leadership can effectively influence front-line employees' emotional labor from an intrinsic motivation perspective. The findings of this study uncover the "black box" between transformational leadership and emotional labor by introducing psychological empowerment as a mediator. This study supplements Luo and Guchait's work [13], which merely examines the direct effect of transformational leadership on emotional labor.

Secondly, the results of this study contribute to the literature on transformational leadership as well. Despite an array of prior studies that have demonstrated that transformational leadership is an effective leadership style for bootstring proactive employee behaviors, such as OCB [81] and innovation behaviors [58,60], only a few studies have recently turned their interests to examine transformational leadership's effects on front-line employees' emotional labor [13], which is a very common form of proactive behavior during their daily work. Not only does this study confirm transformational leadership's effects on front-line employees' emotional labor in the context of service, but this study also illustrates the psychological process of these effects by examining psychological empowerment's mediating effects. These findings answer Siangchokyoo's [82] call for more detailed studies to examine the role of empowerment plays during the link of transformational leadership and employees' behavior.

Thirdly, this study extends research on psychological empowerment. Although prior studies have confirmed that psychological empowerment, as an important psychological construct, is a typical consequence of transformational leadership [82], only a few studies have further extended this link to employees' behavior. The results of this study confirm the existence of the logical path: transformational leadership–psychological empowerment–emotional labor. Furthermore, by considering psychological empowerment as a four-dimension construct (meaning, impact, self-determination, and self-efficacy), this study distinguishes the different roles of four dimensions of psychological empowerment play during the link of transformational leadership and different emotional labor strategies. It therefore pushes forward our understanding of the distinctiveness of four psychological empowerment dimensions.

5.2. Managerial Implications

As competition becomes fiercer [83], service managers are increasingly highlighting the emotions front-line employees display during their interaction with customers. However, given the discretionary nature of front-line employees' emotional labor, it is a big challenge for managers to effectively influence their subordinates' emotional labor. The results of this study may present interesting insights for managers in intervening in their subordinates' emotional labor.

Firstly, service firms have to realize the importance of a leader in influencing front-line employees' emotional labor. According to the findings of this study, transformational leadership is an ideal leadership style for front-line employees. Therefore, during the leader recruiting or promotion process, it is necessary to set criteria considering candidates' competence or personality, which make them easier to perform transformational leadership behaviors. By doing so, service firms can guarantee that leaders are suitable for their job and can effectively exert influence on front-line employees.

Secondly, service managers should alter their leadership style to transformational leadership, which has been suggested to promote front-line employees' deep acting and reduce surface acting. Therefore, training programs for service managers should focus on the skills of engaging transformational leadership behaviors. According to the definition of transformational leadership, typical transformational leadership behaviors include idealized influence, intellectual stimulation, inspirational motivation, and individualized consideration [41]. Once leaders take these behaviors in their routine work, the front-line employees tend to take deep acting rather than surface acting during service encounters. Furthermore, organizational culture also can foster a climate suitable for transformational leadership. If the organizational culture encourages employee self-growth, providing employee future vision, and caring for employee needs and well-being, the leaders are prone to take transformational leadership to comply with the organizational culture.

Thirdly, efforts should be devoted to enhancing front-line employees' psychological empowerment. The results of this study show that psychological empowerment plays an important role in the link of transformational leadership and emotional labor. In order to enhance front-line employees' psychological empowerment, service firms should especially improve employees' sense of impact. For example, service firms can send signals of caring and valuating employees' contribution by providing positive feedback, such as rewarding. To improve front-line employees' self-determination, managers can invite front-line employees to take part in the decision-making. Additionally, enhancing front-line employees' job autonomy through empowerment is an effective way to improve front-line employees' self-determination. In addition, enhancing self-efficacy is necessary to eliminate front-line employees' surface acting. Managers should encourage employees to set challenging goals and provide support to help them to overcome difficulties and complete the tasks.

5.3. Limitations and Future Research

Although this study makes theoretical and practical contributions, there are also several limitations. First, the research context of this study focusses on call centers. Even though emotional labor is very common for call center employees, employees' emotional labor strategies or efforts may be different for other service industries, such as hotel or retail. Therefore, the generalizability of the findings is limited. Future studies should broaden the research contexts to cross industries.

Second, this study has proposed and empirically tested the relationship of transformational leadership, psychological empowerment, and emotional labor. However, the boundary conditions of this relationship are neglected. For example, recent research has demonstrated that the relationship between transformational leadership and psychological empowerment is moderated by organizational structures [84]. Thus, future studies should explore boundary conditions that may moderate the relationship between transformational leadership, psychological empowerment, and emotional labor.

Third, given that the findings of this study suggest a partial mediation between transformational leadership and deep acting, future research should consider other alternative mediating mechanisms. For example, because transformational leaders are more likely to encourage their followers to engage job crafting to enhance their performance [85], it is reasonable to expect that job crafting might mediate the relationship between transformational leadership and deep acting. Therefore, future studies should extend the current findings and explore other potential mediators.

6. Conclusions

Drawing on the conservation of resources (COR) theory, the current study has proposed and tested transformational leadership's effects on the emotional labor of front-line employees via psychological empowerment. These hypotheses were tested using data of 436 front-line employees from call centers. The results reveal that transformational leadership shows positive and negative effects on deep acting and surface acting, respectively. Psychological empowerment exerts mediating effects on the relationship between transformational leadership and front-line employees' emotional labor. Specifically, three dimensions of psychological empowerment (impact, self-determination, self-efficacy) partially mediate transformational leadership's effects on deep acting. Two dimensions of psychological empowerment (impact, self-efficacy) fully mediate the relationship between transformational leadership and surface acting.

Author Contributions: Conceptualization, P.C.; data curation, Z.L.; formal analysis, P.C.; funding acquisition, P.C.; investigation, P.C.; methodology, Z.L.; project administration, P.C.; resources, P.C.; software, Z.L.; supervision, P.C.; validation, L.Z.; visualization, L.Z.; writing—original draft, Z.L. All authors have read and agreed to the published version of the manuscript.

Funding: This research was funded by the Soft Science Research Program of the Shaanxi Province of China (NO.2021KRM157) and the Scientific Research Program Funded by the Shaanxi Provincial Education Department (Program No.21JZ041).

Institutional Review Board Statement: Ethical review and approval were waived for this study because our research did not entail clinical trials of human or animals, and the questionnaire was formed to investigate front-line employees' opinions on the motivations which drive them to engage in different emotional labor strategies during service encounters, which did not encompass ethical issues.

Informed Consent Statement: Informed consent was obtained from all subjects involved in the study.

Data Availability Statement: Not applicable.

Conflicts of Interest: The authors declare no conflict of interest.

References

1. Wen, J.; Huang, S.; Hou, P. Emotional intelligence, emotional labor, perceived organizational support, and job satisfaction: A moderated mediation model. *Int. J. Hosp. Manag.* **2019**, *81*, 120–130. [CrossRef]
2. Hochschild, A.R. *The Managed Heart: Commercialization of Human Feeling*; University of California Press: Los Angeles, CA, USA, 1983.
3. Brotheridge, C.M.; Grandey, A.A. Emotional Labor and Burnout: Comparing Two Perspectives of "People Work". *J. Vocat. Behav.* **2002**, *60*, 17–39. [CrossRef]
4. Brotheridge, C.M.; Lee, R.T. Testing a conservation of resources model of the dynamics of emotional labor. *J. Occup. Health Psychol.* **2002**, *7*, 57–67. [CrossRef] [PubMed]
5. Grandey, A.A. Emotion regulation in the workplace: A new way to conceptualize emotional labor. *J. Occup. Health Psychol.* **2000**, *5*, 85–110. [CrossRef] [PubMed]
6. Grandey, A.A.; Fisk, G.M.; Mattila, A.S.; Jansen, K.J.; Sideman, L.A. Is "service with a smile" enough? Authenticity of positive displays during service encounters. *Organ. Behav. Hum. Decis. Process.* **2005**, *96*, 38–55. [CrossRef]
7. Cheng, B.; Dong, Y.; Zhou, X.; Guo, G.; Peng, Y. Does customer incivility undermine employees' service performance? *Int. J. Hosp. Manag.* **2020**, *89*, 102544. [CrossRef]
8. Yoo, J. Perceived customer participation and work engagement: The path through emotional labor. *Int. J. Bank Mark.* **2016**, *34*, 1009–1024. [CrossRef]
9. Kammeyer-Mueller, J.D.; Rubenstein, A.L.; Long, D.M.; Odio, M.A.; Buckman, B.R.; Zhang, Y.; Halvorsen-Ganepola, M.D.K. A Meta-Analytic Structural Model of Dispositonal Affectivity and Emotional Labor. *Pers. Psychol.* **2013**, *66*, 47–90. [CrossRef]
10. Kim, T.-Y.; Gilbreath, B.; David, E.M.; Kim, S.-P. Self-verification striving and employee outcomes: The mediating effects of emotional labor of South Korean employees. *Int. J. Contemp. Hosp. Manag.* **2019**, *31*, 2845–2861. [CrossRef]
11. Lee, Y.H.; Chelladurai, P. Affectivity, Emotional Labor, Emotional Exhaustion, and Emotional Intelligence in Coaching. *J. Appl. Sport Psychol.* **2016**, *28*, 170–184. [CrossRef]
12. Lu, J.; Zhang, Z.; Jia, M. Does Servant Leadership Affect Employees' Emotional Labor? A Social Information-Processing Perspective. *J. Bus. Ethics* **2019**, *159*, 507–518. [CrossRef]
13. Luo, A.; Guchait, P.; Lee, L.; Madera, J.M. Transformational leadership and service recovery performance: The mediating effect of emotional labor and the influence of culture. *Int. J. Hosp. Manag.* **2019**, *77*, 31–39. [CrossRef]
14. Pieterse, A.N.; Knippenberg, D.V.; Schippers, M.; Stam, D. Transformational and transactional leadership and innovative behaviour: The moderating role of psychological empowerment. *J. Organ. Behav.* **2010**, *31*, 609–623. [CrossRef]
15. Liang, H.-Y.; Tang, F.-I.; Wang, T.-F.; Lin, K.-C.; Yu, S. Nurse characteristics, leadership, safety climate, emotional labour and intention to stay for nurses: A structural equation modelling approach. *J. Adv. Nurs.* **2016**, *72*, 3068–3080. [CrossRef] [PubMed]
16. Aydogmus, C.; Metin Camgoz, S.; Ergeneli, A.; Tayfur Ekmekci, O. Perceptions of transformational leadership and job satisfaction: The roles of personality traits and psychological empowerment. *J. Manag. Organ.* **2018**, *24*, 81–107. [CrossRef]
17. Huang, J. The relationship between employee psychological empowerment and proactive behavior: Self-efficacy as mediator. *Soc. Behav. Personal.* **2017**, *45*, 1157–1166. [CrossRef]
18. Saira, S.; Mansoor, S.; Ali, M. Transformational leadership and employee outcomes: The mediating role of psychological empowerment. *Leadersh. Organ. Dev. J.* **2021**, *42*, 130–143. [CrossRef]
19. Schermuly, C.C.; Meyer, B. Transformational leadership, psychological empowerment, and flow at work. *Eur. J. Work. Organ. Psychol.* **2020**, *29*, 740–752. [CrossRef]
20. Jauhari, H.; Singh, S.; Kumar, M. How does transformational leadership influence proactive customer service behavior of frontline service employees? Examining the mediating roles of psychological empowerment and affective commitment. *J. Enterp. Inf. Manag.* **2017**, *30*, 30–48. [CrossRef]
21. Stanescu, D.F.; Zbuchea, A.; Pinzaru, F. Transformational leadership and innovative work behaviour: The mediating role of psychological empowerment. *Kybernetes* **2021**, *50*, 1041–1057. [CrossRef]
22. Hancer, M.; George, R.T. Psychological empowerment of non-supervisory employees working in full-service restaurants. *Int. J. Hosp. Manag.* **2003**, *22*, 3–16. [CrossRef]
23. Dust, S.B.; Resick, C.J.; Margolis, J.A.; Mawritz, M.B.; Greenbaum, R.L. Ethical leadership and employee success: Examining the roles of psychological empowerment and emotional exhaustion. *Leadersh. Q.* **2018**, *29*, 570–583. [CrossRef]
24. El-Gazar, H.E.; Zoromba, M.A.; Zakaria, A.M.; Abualruz, H.; Abousoliman, A.D. Effect of humble leadership on proactive work behaviour: The mediating role of psychological empowerment among nurses. *Nurs. Manag.* **2022**, *30*, 2689–2698. [CrossRef] [PubMed]
25. Diefendorff, J.M.; Gosserand, R.H. Understanding the emotional labor process: A control theory perspective. *J. Organ. Behav.* **2003**, *24*, 945–959. [CrossRef]
26. Grandey, A.A.; Melloy, R.C. The state of the heart: Emotional labor as emotion regulation reviewed and revised. *J. Occup. Health Psychol.* **2017**, *22*, 407–422. [CrossRef]

27. Ashforth, B.E.; Humphrey, R.H. Emotional Labor in Service Roles: The Influence of Identity. *Acad. Manag. Rev.* **1993**, *18*, 88–115. [CrossRef]
28. Ybema, J.F.; Van Dam, K. The importance of emotional display rules for employee well-being: A multi-group comparison. *J. Posit. Psychol.* **2014**, *9*, 366–376. [CrossRef]
29. Jiun-Lan, H. Effects of Emotional Labor on Organizational Performance in Service Industry. *Pak. J. Stat.* **2012**, *28*, 757–765. [CrossRef]
30. Chi, N.-W.; Grandey, A.A. Emotional labor predicts service performance depending on activation and inhibition regulatory fit. *J. Manag.* **2019**, *45*, 673–700. [CrossRef]
31. Choi, H.-M.; Mohammad, A.A.A.; Kim, W.G. Understanding hotel frontline employees' emotional intelligence, emotional labor, job stress, coping strategies and burnout. *Int. J. Hosp. Manag.* **2019**, *82*, 199–208. [CrossRef]
32. Hur, W.-M.; Moon, T.-W.; Jung, Y.S. Customer response to employee emotional labor: The structural relationship between emotional labor, job satisfaction, and customer satisfaction. *J. Serv. Mark.* **2015**, *29*, 71–80. [CrossRef]
33. Lee, L.; Madera, J.M. Faking it or feeling it: The emotional displays of surface and deep acting on stress and engagement. *Int. J. Contemp. Hosp. Manag.* **2019**, *31*, 1744–1762. [CrossRef]
34. Philipp, A.; Schupbach, H. Longitudinal effects of emotional labour on emotional exhaustion and dedication of teachers. *J. Occup. Health Psychol.* **2010**, *15*, 494–504. [CrossRef] [PubMed]
35. Chi, N.-W.; Grandey, A.A.; Diamond, J.A.; Krimmel, K.R. Want a Tip? Service Performance as a Function of Emotion Regulation and Extraversion. *J. Appl. Psychol.* **2011**, *96*, 1337–1346. [CrossRef] [PubMed]
36. Pugh, S.D.; Diefendorff, J.M.; Moran, C.M. Emotional labor: Organization-level influences, strategies, and outcomes. In *Emotional Labor in the 21st Century: Diverse Perspectives on Emotion Regulation at Work*; Grandey, A., Diefendorff, J., Rupp, D.E., Eds.; Routledge/Taylor & Francis Group: London, UK, 2013; pp. 199–221.
37. Hur, W.-M.; Rhee, S.-Y.; Ahn, K.-H. Positive psychological capital and emotional labor in Korea: The job demands-resources approach. *Int. J. Hum. Resour. Manag.* **2016**, *27*, 477–500. [CrossRef]
38. Chi, N.-W.; Chen, Y.-C.; Huang, T.-C.; Chen, S.-F. Trickle-Down Effects of Positive and Negative Supervisor Behaviors on Service Performance: The Roles of Employee Emotional Labor and Perceived Supervisor Power. *Hum. Perform.* **2018**, *31*, 55–75. [CrossRef]
39. Bass, B.M. From transactional to transformational leadership: Learning to share the vision. *Organ. Dyn.* **1990**, *18*, 19–31. [CrossRef]
40. Avolio, B.J.; Bass, B.M.; Jung, D.I. Re-examining the components of transformational and transactional leadership using the Multifactor Leadership Questionnaire. *J. Occup. Organ. Psychol.* **1999**, *72*, 441–462. [CrossRef]
41. Bass, B.M.; Riggio, R.E. *Transformational Leadership*, 2nd ed.; Lawrence Erlbaum Associates Publishers: Mahwah, NJ, USA, 2006.
42. Avolio, B.J.; Zhu, W.; Koh, W.; Bhatia, P. Transformational leadership and organizational commitment: Mediating role of psychological empowerment and moderating role of structural distance. *J. Organ. Behav.* **2004**, *25*, 951–968. [CrossRef]
43. Choi, S.L.; Goh, C.F.; Adam, M.B.H.; Tan, O.K. Transformational leadership, empowerment, and job satisfaction: The mediating role of employee empowerment. *Hum. Resour. Health* **2016**, *14*, 1–14. [CrossRef]
44. Elshout, R.; Scherp, E.; Feltz-Cornelis, C.M.V.D. Understanding the link between leadership style, employee satisfaction, and absenteeism: A mixed methods design study in a mental health care institution. *Neuropsychiatr. Dis. Treat.* **2013**, *9*, 823–837. [CrossRef] [PubMed]
45. Mahalinga Shiva, M.; Suar, D. Transformational Leadership, Organizational Culture, Organizational Effectiveness, and Programme Outcomes in Non-Governmental Organizations. *Int. J. Volunt. Nonprofit Organ.* **2012**, *23*, 684–710. [CrossRef]
46. Jaskyte, K. Transformational Leadership, Organizational Culture, and Innovativeness in Nonprofit Organizations. *Nonprofit Manag. Leadersh.* **2004**, *15*, 153–168. [CrossRef]
47. Lasrado, F.; Kassem, R. Let's get everyone involved! The effects of transformational leadership and organizational culture on organizational excellence. *Int. J. Qual. Reliab. Manag.* **2021**, *38*, 169–194. [CrossRef]
48. Bass, B.M. *Transformational Leadership and Team and Organizational Decision Making*; Sage: Thousand Oaks, CA, USA, 1994.
49. Bass, B.M. Two decades of research and development on transformational leadership. *Eur. J. Work. Organ. Psychol.* **1999**, *8*, 9–32. [CrossRef]
50. Walumbwa, F.O.; Hartnell, C.A.; Oke, A. Servant Leadership, Procedural Justice Climate, Service Climate, Employee Attitudes, and Organizational Citizenship Behavior: A Cross-Level Investigation. *J. Appl. Psychol.* **2010**, *95*, 517–529. [CrossRef]
51. Bose, S.; Patnaik, B.; Mohanty, S. The Mediating Role of Psychological Empowerment in the Relationship Between Transformational Leadership and Organizational Identification of Employees. *J. Appl. Behav. Sci.* **2020**, *57*, 490–510. [CrossRef]
52. Dutu, R.; Butucescu, A. On the Link between Transformational Leadership and Employees' Work Engagement: The Role of Psychological Empowerment. *Hum. Resour. Psychol.* **2019**, *17*, 76–87. [CrossRef]
53. Richardson, H.A.; Kluemper, D.H.; Taylor, S.G. Too little and too much authority sharing: Differential relationships with psychological empowerment and in-role and extra-role performance. *J. Organ. Behav.* **2021**, *42*, 1099–1119. [CrossRef]
54. Hackman, J.R. Group influences on individuals in organizations. In *Handbook of Industrial and Organizational Psychology*; Dunnette, M.D., Hough, L.M., Eds.; Consulting Psychologists Press: Palo Alto, CA, USA, 1992; Volume 3, pp. 199–267.
55. Thomas, K.W.; Velthouse, B.A. Cognitive Elements of Empowerment: An "Interpretive" Model of Intrinsic Task Motivation. *Acad. Manag. Rev.* **1990**, *15*, 666–681. [CrossRef]
56. Spreitzer, G.M. Psychological empowerment in the workplace: Dimensions, measurement, and validation. *Acad. Manag. J.* **1995**, *38*, 1442–1465. [CrossRef]

57. Ryan, R.M.; Deci, E.L. Self-determination theory and the facilitation of intrinsic motivation, social development, and well-being. *Am. Psychol.* **2000**, *55*, 68–78. [CrossRef] [PubMed]

58. Gumusluoglu, L.; Ilsev, A. Transformational leadership, creativity, and organizational innovation. *J. Bus. Res.* **2009**, *62*, 461–473. [CrossRef]

59. Hobfoll, S.E.; Halbesleben, J.; Neveu, J.-P.; Westman, M. Conservation of resources in the organizational context: The reality of resources and their consequences. *Annu. Rev. Organ. Psychol. Organ. Behav.* **2018**, *5*, 103–128. [CrossRef]

60. Grošelj, M.; Černe, M.; Penger, S.; Grah, B. Authentic and transformational leadership and innovative work behaviour: The moderating role of psychological empowerment. *Eur. J. Innov. Manag.* **2020**, *24*, 677–706. [CrossRef]

61. Bandura, A. *Self-Efficacy: The Exercise of Control*; Freeman: New York, NY, USA, 1997.

62. Grandey, A.A.; Dickter, D.N.; Sin, H.-P. The customer is not always right: Customer aggression and emotion regulation of service employees. *J. Organ. Behav.* **2004**, *25*, 397–418. [CrossRef]

63. Amabile, T.M.; Conti, R.; Coon, H.; Lazenby, J.; Herron, M. Assessing the work environment for creativity. *Acad. Manag. J.* **1996**, *39*, 1154–1184. [CrossRef]

64. Bakker, A.B.; Demerouti, E. Job demands–resources theory: Taking stock and looking forward. *J. Occup. Health Psychol.* **2017**, *22*, 273–285. [CrossRef]

65. Grandey, A.A.; Sayre, G.M. Emotional Labor: Regulating Emotions for a Wage. *Curr. Dir. Psychol. Sci.* **2019**, *28*, 131–137. [CrossRef]

66. McGraw, K.; Fiala, J. Undermining the Zeigarnik effect: Another hidden cost of reward. *J. Personal.* **1982**, *50*, 58–66. [CrossRef]

67. Ashforth, B.E. The experience of powerlessness in organizations. *Organ. Behav. Hum. Decis. Process.* **1989**, *43*, 207–242. [CrossRef]

68. Deci, E.L.; Connell, J.P.; Ryan, R.M. Self–determination in a work organization. *J. Appl. Psychol.* **1989**, *74*, 580–590. [CrossRef]

69. Bell, N.E.; Staw, B.M. People as sculptors versus sculpture: The roles of personality and personal control in organizations. In *Handbook of Career Theory*; Cambridge University Press: Cambridge, UK, 1989; pp. 232–251. [CrossRef]

70. Liden, R.C.; Wayne, S.J.; Sparrowe, R.T. An examination of the mediating role of psychological empowerment on the relations between the job, interpersonal relationships, and work outcomes. *J. Appl. Psychol.* **2000**, *85*, 407–416. [CrossRef] [PubMed]

71. Seibert, S.E.; Wang, G.; Courtright, S.H. Antecedents and consequences of psychological and team empowerment in organizations: A meta-analytic review. *J. Appl. Psychol.* **2011**, *96*, 981–1003. [CrossRef]

72. Bandura, A. *Social Foundations of Thought and Action: A Social Cognitive View*; Prentice-Hall: Hoboken, NJ, USA, 1987.

73. Gist, M. Self–efficacy: Implications for organizational behavior and human resource management. *Acad. Manag. Rev.* **1987**, *12*, 472–485. [CrossRef]

74. Luszczynska, A.; Scholz, U.; Schwarzer, R. The general self-efficacy scale: Multicultural validation studies. *J. Psychol.* **2005**, *139*, 439–457. [CrossRef]

75. Judge, T.A.; Bono, J.E.; Erez, A.; Locke, E.A. Core self-evaluations and job and life satisfaction: The role of self-concordance and goal attainment. *J. Appl. Psychol.* **2005**, *90*, 257–268. [CrossRef]

76. Abraham, R. Emotional dissonance in organizations: Conceptualizing the roles of self-esteem and job-induced tension. *Leadersh. Organ. Dev. J.* **1999**, *20*, 18–25. [CrossRef]

77. Brislin, R.W. Translation and content analysis of oral and written material. In *Handbook of Cross-Cultural Psychology: Methodology*; Triandis, H.C., Berry, J.W., Eds.; Allyn and Bacon: Boston, MA, USA, 1980; Volume 2, pp. 349–444.

78. Brotheridge, C.l.M.; Lee, R.T. Development and validation of the Emotional Labour Scale. *J. Occup. Organ. Psychol.* **2010**, *76*, 365–379. [CrossRef]

79. Podsakoff, P.M.; MacKenzie, S.B.; Lee, J.Y.; Podsakoff, N.P. Common method biases in behavioral research: A critical review of the literature and recommended remedies. *J. Appl. Psychol.* **2003**, *88*, 879–903. [CrossRef]

80. Hayes, A.F. *Introduction to Mediation, Moderation, and Conditional Process Analysis: A Regression-Based Approach*; Guilford Publications: New York, NY, USA, 2017.

81. Purwanto, A.; Purba, J.T.; Bernarto, I.; Sijabat, R. Effect of transformational leadership, job satisfaction, and organizational commitments on organizational citizenship behavior. *Inovbiz J. Inov. Bisnis* **2021**, *9*, 61–69. [CrossRef]

82. Siangchokyoo, N.; Klinger, R.L.; Campion, E.D. Follower transformation as the linchpin of transformational leadership theory: A systematic review and future research agenda. *Leadersh. Q.* **2020**, *31*, 101341. [CrossRef]

83. Noh, H.; Song, Y.; Park, A.-S.; Yoon, B.; Lee, S. Development of new technology-based services. *Serv. Ind. J.* **2016**, *36*, 200–222. [CrossRef]

84. Dust, S.B.; Resick, C.J.; Mawritz, M.B. Transformational leadership, psychological empowerment, and the moderating role of mechanistic–organic contexts. *J. Organ. Behav.* **2014**, *35*, 413–433. [CrossRef]

85. Hetland, J.; Hetland, H.; Bakker, A.B.; Demerouti, E. Daily transformational leadership and employee job crafting: The role of promotion focus. *Eur. Manag. J.* **2018**, *36*, 746–756. [CrossRef]

International Journal of
Environmental Research and Public Health

MDPI

Article

The Influence of Perceived External Prestige on Emotional Labor of Frontline Employees: The Mediating Roles of Organizational Identification and Impression Management Motive

Pengfei Cheng * , **Jingxuan Jiang and Zhuangzi Liu**

School of Economics and Management, Xi'an University of Technology, Xi'an 710054, China
* Correspondence: pfcheng@xaut.edu.cn

Abstract: Drawing on both the organization identification and impression management theories, we propose that perceived external prestige of frontline employees influences their emotional labor through organizational identification and impression management motive. Further, the relative influence of either pathway depends upon perceived organizational support. Using survey data from 377 frontline employees in 104 hotels, the results indicate that perceived external prestige is positively related to deep acting, and negatively related to surface acting. Organizational identification partially mediates the relationship between perceived external prestige and deep acting. However, the relationship between perceived external prestige and surface acting is partially mediated both by organizational identification and impression management motive. In addition, perceived organizational support positively moderates the relationship between perceived external prestige and organizational identification, and negatively moderates the relationship between perceived external prestige and impression management motive, respectively.

Keywords: perceived external prestige; emotional labor; organization identification; impression management motive; perceived organizational support

Citation: Cheng, P.; Jiang, J.; Liu, Z. The Influence of Perceived External Prestige on Emotional Labor of Frontline Employees: The Mediating Roles of Organizational Identification and Impression Management Motive. *Int. J. Environ. Res. Public Health* **2022**, *19*, 10778. https://doi.org/10.3390/ijerph191710778

Academic Editor: Masahito Fushimi

Received: 5 August 2022
Accepted: 27 August 2022
Published: 30 August 2022

Publisher's Note: MDPI stays neutral with regard to jurisdictional claims in published maps and institutional affiliations.

1. Introduction

As competition becomes fiercer, service firms rely more on providing an excellent service experience to survive. During service encounters, how frontline employees regulate and display emotions is so important in shaping customers' service experience that almost all service firms take "service with a smile" as a very important mantra [1]. Hochschild [2] defined emotional labor as how employees regulate and display their emotions during service encounters to conform with organizations' display rules. Although there are two different strategies employees can take to engage in emotional labor, surface acting and deep acting [2,3], the outcomes of deep acting and surface acting are different. Deep acting refers to employees trying to display appropriate emotions by actually experiencing or feeling the emotions; it contributes to positive outcomes such as customer emotional experience [4], employee job satisfaction [5], and perceived service quality [6]. Surface acting refers to employees feigning emotions that are not actually felt; it causes negative outcomes such as burnout [7], negative customer emotion [8], and negatively affecting customers' subjective perception [9].

Given the pivotal role of emotional labor in shaping service experience [6], both academics and practitioners are increasingly interested in how to effectively manage frontline employees' emotional labor. However, considering the dynamism and complexity of service interactions, it is impossible for managers to monitor whether an employee engaged in deep or surface acting during the service encounters. Frontline employees' emotional labor, to a great extent discretionary behaviors [10], goes beyond their supervisor's direct control and are mainly driven by employees' intrinsic motivation [11]. Hence, it is important for service firms to explore factors that can serve as intrinsic motivations of frontline employees' emotional labor and understand the mechanisms by which the influence was exerted.

Smidts, et al. [12] defined perceived external prestige (PEP) as an employee's perception of how outsiders judge the status and image of his or her organization. According to social identity theory, an individual may generate strong attachment, pride, and self-enhancement motive to his or her organization that holds great reputation [13]. Previous research suggests that PEP plays a pivotal role in employees' attitudes and behavior, especially their discretionary actions (e.g., organizational citizenship behavior) [13,14]. Yet there is little research regarding the influence of PEP on the emotional labor of frontline employees [15]. In addition, as an important organizational phenomenon, impression management has a critical influence on employees and their firms. For example, impression management motive fosters employees' organizational citizenship behavior and positivity [16,17]. Emotional display, as one type of employee impression management behavior [18], can be motivated by impression management motivation. However, researchers to date have paid scant attention to the effect of PEP on employees' behavior from an impression management perspective. Therefore, the purpose of this research is to explore the relationship between PEP, impression management motive, organizational identification, and employees' emotional labor.

Drawing upon the impression management theory and social identity theory, we propose that PEP influences employees' emotional labor through dual mediating processes, namely organizational identification and impression management motive. Furthermore, perceived organizational support (POS) affects employees' beliefs concerning their legitimacy as organizational members [19,20]. The external prestige of the organization is salient to employees only when they feel themselves to be legitimate organizational members [21]. Thus, we also highlight the moderating role of POS, which may alter the relationship between PEP and employees' emotional labor. This research may extend previous literature on emotional labor and perceived external prestige by opening the "black box" between PEP and emotional labor and considering the boundary conditions of these relationships.

2. Theoretical Background and Hypothesis Development

2.1. Emotional Labor

Emotional labor is defined as behaviors by frontline employees for displaying appropriate emotions to conform to organizational display rules [2]. Previous studies have described two dimensions of emotional labor strategies: surface acting and deep acting [2,3]. Specifically, surface acting involves a faking process, where employees express the expected emotion without necessarily altering how they feel. In contrast, deep acting is described as a more effortful process, where employees try to display expected emotions by adjusting their inner feelings [3]. Due to the importance of emotional labor in the service experience, in recent years, considerable research has tried to help service firms control and manage frontline employees' emotional labor by identifying antecedents from an organizational perspective, such as organizational justice [22], organizational dehumanization [23], and leadership [24]. As a discretionary behavior, frontline employees' emotional labor is prone to be driven by their intrinsic motivations. However, the mechanisms by which these organizational factors internalize and influence frontline employees' emotional labor have received little explicit attention in organizational scholarship. Hence, addressing the antecedents to emotional labor from an organizational perspective and unveiling the process of these effects may enable service firms to manage employees' emotional labor more effectively, and consequently create excellent service experience.

2.2. Perceived External Prestige

Perceived external prestige (PEP) refers to the employee's individual beliefs about how the external relevant stakeholders, such as customers, competitors, and suppliers, view his/her organization [12,25]. It is important to note that external prestige and organizational reputation are closely associated, but two distinct constructs. Organizational reputation refers to the overall assessment of current organizational assets, market position, and future behavior [26], which reflects outsiders' beliefs. In contrast, PEP reflects insiders' (e.g., employees) beliefs about their organization. For the same organization, the cognitions or evaluations of outsiders and insiders may result from different sources of information [12]. Thus, even though an organization's external prestige is often related to its reputation, they are two different constructs.

2.2.1. The Mediating Role of Organizational Identification

Prior research on external prestige suggests that the effects of PEP on employees' attitudes and behavior toward their organization are mediated by organizational identification [12]. As a psychological foundation of the employee–organization relationship, organizational identification refers to the extent that an employee's perception of oneness with or belonging to her or his organization [21]. PEP reflects how the public view the organization, and is acknowledged as an antecedent of organizational identification [27]. Driven by employees' self-enhancement motive, high external prestige not only enhances the attractiveness of organizational membership to an employee but ultimately results in proactive behaviors [28]. During service encounters, frontline employees, who act as representatives of the service firm in interactions with customers, are likely to depersonalize the self and use the organization as a vehicle for self-definition [29]. Under this condition, organizations with high external prestige can enhance or maintain employees' positive social identity [30]. Thus, PEP contributes to enhancing employees' self-esteem and organization identification.

Hypothesis 1. *Perceived external prestige is positively related to organizational identification.*

Given the discretionary nature of frontline employees' emotional labor [10], it is logical to expect that frontline employees' deep acting or surface acting during the service encounter is driven by their intrinsic motivation (e.g., organizational identification) [31]. According to social identity theory, employees who strongly identify with their firms are prone to take organizational goals as their own and consequently engage in behaviors that are beneficial to the firms [32]. For instance, employees with high organizational identification are more likely to engage in OCB [33] and customer orientation behaviors [34], which are especially important in service industries. Therefore, we can expect that, during service encounters, frontline employees with high organizational identification would internalize organizational display rules and devote more effort to altering their inner feelings to obey the display rules, namely engaging in deep acting [35]. Conversely, employees with low organizational identification would ignore organizational interests and display rules by merely pretending "fake" emotions, namely conducting surface acting.

Hypothesis 2. *Organizational identification is positively related to deep acting.*

Hypothesis 3. *Organizational identification is negatively related to surface acting.*

2.2.2. The Mediating Role of the Impression Management Motive

Impression management theory suggests that people care about how they are viewed by others [16,36]. To shape a new or maintain a current personal image, employees manage their image that is projected to the target population (e.g., interviewers, supervisors, or customers) through strategic behavior at work [37,38]. Impression management processes are generally conscious and tactical because employees are prone to shape or maintain specific images [39].

Research on impression management has typically focused on the internal context of organizations, such as how employees affect the personal perceptions of their colleagues, supervisors, and subordinates of them through impression management [16,40,41]. Although the targets of employees' impression management also include their customers [42], few studies pay attention to frontline employees' impression management when interacting with customers. Yun et al. [43] defined an individual's impression management motive as the desire to enhance one's self-image by consciously exhibiting specific behaviors. When frontline employees, as a service firm's representatives, interact with customers, the external prestige of the employees' organization becomes an important and valued component of their self-image. Therefore, frontline employees have strong motive to use impression management strategies to maintain their positive self-image, which is derived from their organization's prestige. Thus, we propose:

Hypothesis 4. *Perceived external prestige is positively related to impression management motive.*

As representatives of service firms, frontline employees deliberately attempt to regulate and display their emotions (emotional labor) during the service encounter, in order to shape customers' perception of themselves and the firms. Therefore, Ashforth and Humphrey [44] argue that emotional labor can be considered a form of impression management. There are many common skills which can be used for both emotional labor and impression management. Bolino, Long, and Turnley [16] indicate that smiling and eye contact may be both important impression management tactics and salient ways for employees to display emotions during service interactions. Verbal greetings and farewells not only are an important way for frontline employees to express positive emotions [16], but also symbolize employees' willingness to build, maintain, and enhance customer relationships, which is similar to ingratiation tactics in impression management. Thus, we can expect that frontline employees' emotional labor would be driven by an impression management motive.

In general, people's impression management is conscious and tactical [39]. Therefore, previous research has suggested that the image a person projects to others through impression management behaviors is perhaps not authentic, but rather fake [38], which is a similarity between impression management and employee surface acting. Ashforth and Humphrey [44] mentioned that surface acting, which focuses directly on one's outward behavior, was the form of acting typically discussed as impression management. In contrast, deep acting focuses on one's inner feelings, beyond the notion of impression management. In other words, driven by impression management motives, employees' positive displays may be due to maintaining or enhancing the image they project to customers, rather than expressing authentic emotions they feel. We can also get similar statements from Albrecht et al. [45], who argued that surface acting is an important impression management tactic, and Shumski Thomas et al. [46], who found that many employees hide authentic emotions by surface acting to avoid offending their supervisors and maintain a good image. Thus, we propose that employees, driven by impression management motives, are prone to engage in more surface acting and less deep acting during service interactions.

Hypothesis 5. *Impression management motive is negatively related to deep acting.*

Hypothesis 6. *Impression management motive is positively related to surface acting.*

2.3. The Moderating Role of Perceived Organizational Support

People generally are motivated to improve and maintain their status and self-worth [47]. Frontline service employees, as boundary-spanners, interact not only with customers outside the firm, but also with supervisors and colleagues inside the organization. Frontline employees' self-worth therefore depends on both their status during the service encounters and their status in the organization. External prestige reflects the organization's status, from which frontline employees may inform "how others outside the organization view me". Thus, high perceived external prestige (PEP) enables frontline employees to internalize a high status when interacting with customers. Perceived organizational support (POS) refers to the perceptions of employees that their organization values and cares about their contributions and well-being [48,49]. As a sign of employees' status in the organization, POS reflects "how others in the organization (e.g., supervisors or colleagues) view me". A high level of POS enables employees to experience feelings of respect [14,50].

Previous research suggests that POS affects employees' beliefs about the legitimacy of their organizational membership [21,51]. When employees' contributions are valued by an organization, they will perceive they are respected and affectively attached to the organization [48]. Under this condition, employees perceive the legitimacy of their organizational membership. Conversely, if employees perceive less organizational support, they are merely "nominal" rather than legitimate members of the organization. According to social identity theory, employees feel obligated to care about the external status of the organization (external prestige) only when they believe that their organizational membership is legitimate [21,51]. In other words, the external prestige of the firm makes sense and results in employees' identification only when they perceive a high level of organizational support. In contrast, under the condition of low POS, employees might be unconcerned about the external prestige of the organization, which will undermine the relationship between external prestige and organizational identification.

Hypothesis 7. *The positive relationship between perceived external prestige and organizational identification will be moderated by perceived organizational support. Specifically, the relationship will be stronger (weaker) when employees have a high (low) level of perceived organizational support.*

Previous research has demonstrated that PEP and POS are both related to employees' self-worth. However, these two constructs may be inconsistent for an employee. To enhance external prestige, companies send positive information to external stakeholders via marketing communication [52,53]. However, an employee's perception of their organization might be different from outsiders' perception. For example, a firm with high external prestige may surprisingly support its employees less [54]. In this context, although external prestige can make frontline employees feel a high level of self-worth and status when they interact with customers, employees may not generate an attachment to the organization due to the lack of respect and legitimate organizational membership. Thus, frontline employees are more likely to engage in impression management tactics driven by instrumental motivations. In other words, employees with low POS are prone to capture the benefits of PEP (e.g., higher status and self-worth) by deliberately engaging in impression management during the service encounter. Furthermore, employees with low POS doubt the way the organization treats them, and impression management is an effective means for employees to cope with the perception of ambivalence (high external prestige versus low organizational support). Figure 1 presents the research framework.

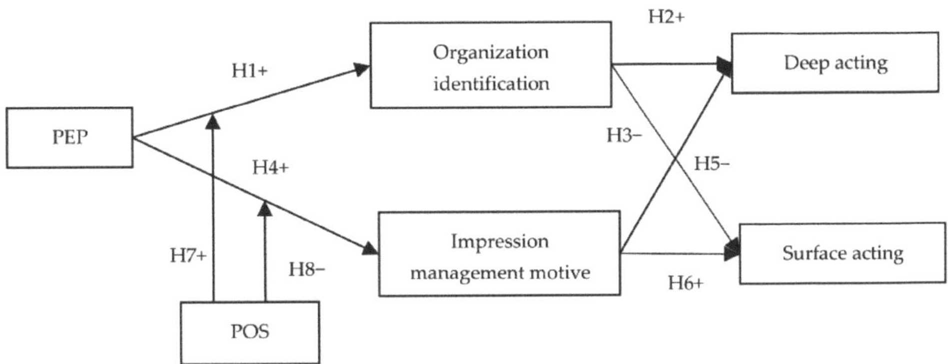

Figure 1. Conceptual model.

Hypothesis 8. *The positive relationship between perceived external prestige and impression management motive will be moderated by perceived organizational support. Specifically, the relationship will be stronger (weaker) when employees perceive low (high) organizational support.*

3. Method

3.1. Sample and Procedure

This research selected frontline employees from 104 Chinese hotels for several reasons. First, in response to fiercer competition in the Chinese hospitality industry, external prestige, as one of the intangible assets, has become an important means for hotels to establish their sustainable competitiveness. Second, various tiers of hotels provide sufficient variation in external prestige for this study. Finally, almost all hotels have emotional requirements for frontline employees who are directly in contact with customers during service encounters, which means emotional labor is very common in frontline hotel employees' daily work.

Questionnaires were sent to 520 frontline employees working at hotels located in Xi'an and Chengdu in China. We selected 80 hotels from four tiers of hotels, with 20 for each tier: economy hotels, three-star hotels, four-star hotels, and five-star hotels. For each hotel, 6–8 frontline employees who directly come in contact with customers were selected to take part in the survey. Finally, a total of 390 questionnaires were collected, with 182 from hotels located in Chengdu and 208 from hotels located in Xi'an, respectively. After deleting 13 incomplete questionnaires, a sample of 377 was used for data analysis, with a response rate of 72.5%. The sample consisted of 27.4% males and 72.6% females.

3.2. Measures

For all measures, except for the demographic variables, seven-point Likert-type scales were used ranging from 1 to 7 with 1 indicating "strongly disagree," and 7 indicating "strongly agree".

Core variables. The emotional labor of frontline employees was measured with two 3-item subscales adapted from Brotheridge and Lee [55]. One 3-item subscale captured surface acting and the other captured deep acting. Sample items include "Hide my true feelings about a situation" and "Make an effort to actually feel the emotions that I need to display to others". Organizational identification was measured with the 6-item scale adapted from Homburg, Wieseke, and Hoyer [34], originally developed by Mael and Ashforth [56]. A sample item is "This hotel's success is my success". The 6-item scale that was developed by Yun, Takeuchi and Liu [43] was used to measure impression management motive in this research. A sample item was "It is important to me to give a good impression to others". Concerning perceived external prestige, we adopted a 6-item scale from Herrbach et al. [57]. A sample item was "My organization is considered one of the best". Perceived organizational support was measured using 6 items by Eisenberger et al. [58]. The items have been used by many studies [59,60].

Control variables. We controlled participants' gender (0 = male; 1 = female), age, and tenure. Furthermore, frontline employees in top-tier hotels need to obey more stringent emotional display rules and pay more attention to their emotional regulation at work. Thus, we also controlled hotel levels: economy hotels, three-star hotels, four-star hotels, and five-star hotels. Three dummy variables were used for coding hotel levels (five-star: level 1 = 0, level 2 = 0, level 3 = 0; four-star: level 1 = 1, level 2 = 0, level 3 = 0; three-star: level 1 = 0, level 2 = 1, level 3 = 0; economy: level 1 = 0, level 2 = 0, level 3 = 1).

3.3. Confirmatory Factor Analysis, Reliability, and Common Method Bias

Confirmatory factor analyses (CFA) were conducted to assess the convergent and discriminant validity of the latent constructs used in this research. The overall fit statistics for hypothesized 6-factor measurement model are acceptable: Chi-square = 1122.13, df = 390, root mean square of approximation (RMSEA) = 0.07, normed fit index (NFI) = 0.93, non-normed fit index (NNFI) = 0.95, comparative fit index (CFI) = 0.95. The factor loadings of all items on their corresponding construct were greater than 0.6 (see Table 1), which indicates the convergent validity was acceptable. For the discriminant validity, as shown in Table 2, the square roots of every construct's average variance extracted values (AVE) were greater than the corresponding correlation between all constructs, indicating the discriminant validity is acceptable. Finally, acceptable reliability was confirmed by the fact that the Cronbach's alpha coefficients of all constructs, ranging from 0.781 to 0.920, were above the cutoff value 0.70 as shown in Table 1.

Table 1. CFA and Reliability.

Variables	Items	Factor Loadings	T	Variables	Items	Factor Loadings	T
Deep acting $\alpha = 0.781$	DA-1	0.68	13.50	Surface acting $\alpha = 0.866$	SA-1	0.81	18.06
	DA-2	0.78	15.98		SA-2	0.84	18.92
	DA-3	0.76	15.65		SA-3	0.83	18.39
OI $\alpha = 0.903$	OI-1	0.78	17.39	IMM $\alpha = 0.908$	IM-1	0.82	19.06
	OI-2	0.79	17.71		IM-2	0.81	18.46
	OI-3	0.79	17.85		IM-3	0.80	18.31
	OI-4	0.83	19.22		IM-4	0.81	18.55
	OI-5	0.81	18.55		IM-5	0.79	17.74
	OI-6	0.69	14.66		IM-6	0.70	15.12
PEP $\alpha = 0.920$	PEP-1	0.78	17.60	POS $\alpha = 0.876$	POS-1	0.73	15.83
	PEP-2	0.81	18.52		POS-2	0.82	18.54
	PEP-3	0.85	20.22		POS-3	0.79	17.77
	PEP-4	0.79	17.95		POS-4	0.83	18.90
	PEP-5	0.83	19.49		POS-5	0.62	12.63
	PEP-6	0.81	18.74		POS-6	0.62	12.82

Note: OI = organization identification; PEP = perceived external prestige; IMM = impression management motive; POS = perceived organizational support.

Table 2. Correlation matrix and AVE.

	Means	SD	1	2	3	4	5	6
1. Deep acting	4.38	1.053	0.741					
2. Surface acting	4.34	1.301	−0.223 **	0.843				
3. PEP	4.26	1.328	0.390 **	−0.113 *	0.812			
4. OI	4.16	1.171	0.413 **	−0.262 **	0.515 **	0.783		
5. IMM	4.06	1.166	0.111 *	0.296 **	0.311 **	0.067	0.789	
6. POS	4.35	1.044	0.415 **	−0.130 *	0.208 **	0.277 **	0.073	0.740

Notes: OI = organization identification; PEP = perceived external prestige; IMM = impression management motive; POS = perceived organizational support. The square roots of AVE are presented in diagonal elements (bold values). * $p < 0.05$; ** $p < 0.01$.

Next, we examined the possible effects of common method bias. A Harman's one-factor test was used to rule out possible common method variance (CMV) problems [61]. The results show that six factors were extracted when eigen-values were above 1, and the first factor accounted for only 27.12% of the total variance, providing the evidence that CMV was not present.

4. Results

4.1. Test of the Mediating Effect

To test the mediation effects of organizational identification and impression management motive on the relationships between perceived external prestige (PEP) and emotional labor, we conducted a series of mediation model analyses using PROCESS 3.3 macro for SPSS developed by Hayes [62]. Based on 5000 bootstrapped resamples, PROCESS can provide 95% confidence intervals (CI) for total effects, direct effects, and indirect effects. If the CI excludes zero, the effect is significant. Therefore, it is suitable for testing mediation effects.

First, we tested the effect of perceived external prestige (PEP) on deep acting (DA) via organizational identification (OI) and impression management motive (IMM) by using a simple mediation model from the PROCESS macro. As shown in Table 3 M-1, PEP was positively related to deep acting ($\beta = 0.404$, $p < 0.001$). We supposed PEP influences deep acting via organizational identification and impression management motive. Thus, we examined the effects of PEP on organizational identification, in M-2 and impression management motive in M-3, respectively. The results indicated that PEP was positively related to organizational identification ($\beta = 0.617$, $p < 0.001$), supporting H1. The results indicate that a firm with a high level of external prestige is more likely to evoke employees' identification, which is in line with previous research on organizational identification. As shown in M-3, PEP was positively related to impression management motive ($\beta = 0.334$, $p < 0.001$) as well, supporting H4. Then, in M-4, both PEP ($\beta = 0.242$, $p < 0.001$) and organizational identification ($\beta = 0.256$, $p < 0.001$) were positively related to deep acting, supporting H2. However, as the relationship between impression management motive and deep acting was not significant ($\beta = 0.014$, $p > 0.05$), H5 was rejected. As shown in Table 4, results of the bootstrapping sample (95% confidence interval) indicated that organization identification (CI = 0.053, 0.179) rather than impression management motive (CI = -0.012, 0.030) mediates the impact of PEP on deep acting. In addition, PEP exerts positive direct effect on deep acting significantly ($\beta = 0.242$, $p < 0.001$) (M-4). Taken together, these results confirm that PEP's effect on deep acting is only partially mediated by organizational identification, rather than by impression management motive.

Table 3. Test of mediating effects.

Independent Variable		M-1 DA	M-2 OI	M-3 IMM	M-4 DA	M-5 SA	M-6 SA
Control variable	Gender	0.322 ***	0.044	0.012	0.311 ***	0.409 ***	0.416 ***
	Age	0.090	0.113	0.153	0.117	0.255 **	0.167 *
	Tenure	−0.122 **	−0.016	0.085	−0.119 **	0.004	−0.033
	Level 1	−0.039	0.003	0.229	−0.043	−0.075	−0.162
	Level 2	−0.066	−0.049	0.016	−0.054	−0.041	−0.059
	Level 3	−0.323 **	−0.145	−0.077	−0.285 *	−0.165	−0.172
PEP		0.404 ***	0.617 ***	0.334 ***	0.242 ***	−0.183 ***	−0.155 **
OI		–	–	–	0.256 ***	–	−0.253 ***
IMM		–	–	–	0.014	–	0.386 ***
R^2		0.194 ***	0.281 ***	0.116 ***	0.252 ***	0.056 ***	0.211 ***

Notes: OI = organization identification; PEP = perceived external prestige; IMM = impression management motive; DA = deep acting; SA = surface acting; * $p < 0.05$; ** $p < 0.01$; *** $p < 0.001$.

Table 4. Total, direct, and indirect effects of PEP on deep acting.

		Effect	*Boot SE*	*Boot LLCI*	*Boot ULLCI*
Total effects		0.305	0.038	0.230	0.379
Direct effects		0.182	0.045	0.094	0.271
Indirect effects	PEP→OI→DA	0.119	0.035	0.053	0.179
	PEP→IMM→DA	0.004	0.013	−0.012	0.030

Notes: PEP = perceived external prestige; OI = organization identification; IMM = impression management motive; DA = deep acting; SA = surface acting.

Second, we tested the effect of perceived external prestige (PEP) on surface acting (SA) via organizational identification (OI) and impression management motive (IMM). As shown in Table 3 M-5, PEP was negatively related to surface acting ($\beta = -0.183$, $p < 0.001$). In M-6, the relationship between organizational identification and surface acting was negative and significant ($\beta = -0.253$, $p < 0.001$), supporting H3. The effect of impression management motive on surface acting was positive and significant ($\beta = 0.386$, $p < 0.001$), supporting H6. After controlling the mediators (organizational identification and impression management motive), PEP significantly exerted a negative effect on surface acting ($\beta = -0.155$, $p < 0.01$). As shown in Table 5, results of the bootstrapping sample (95% confidence interval) indicated that both organization identification (CI = −0.196, −0.042) and impression management motive mediated (CI = 0.051, 0.152) the impact of PEP on surface acting. Taken together, these results confirmed that PEP's effect on surface acting was partially mediated by both organizational identification and impression management motive. A side note of interest is that PEP's indirect effects on surface acting via organizational identification and impression management motive are negative and positive, respectively. We will discuss these inconsistent mediations later.

Table 5. Total, direct, and indirect effects of PEP on surface acting.

		Effect	*Boot SE*	*Boot LLCI*	*Boot ULLCI*
	Total effects	−0.138	0.051	−0.238	−0.038
	Direct effects	−0.117	0.057	−0.230	−0.004
Indirect	PEP→OI→SA	−0.118	0.040	−0.196	−0.042
effects	PEP→IMM→SA	0.097	0.026	0.051	0.152

Notes: PEP = perceived external prestige; OI = organization identification; IMM = impression management motive; DA = deep acting; SA = surface acting.

4.2. Test of the Moderated Mediation

To test the moderating role of perceived organizational support, we conducted moderated mediation analyses using Hayes' PROCESS macro. In line with Aiken et al.'s [63] guidelines for moderated regression, the predictor (PEP) and moderator (POS) variables were mean-centered before creating the interaction term. As shown in Table 6, the interaction of PEP and POS was positively related to organizational identification (β = 0.083, $p < 0.05$), and the R^2 change between M-7 and M-8 was significant (ΔR^2 = 0.008, $p < 0.01$). According to Aiken, West, and Reno [63], we graphed the simple slopes of PEP on organizational identification at high and low level of POS (+1.0, and −1.0 standard deviations from mean), and visualized the form of the moderating effect. As shown in Figure 2, the positive relationship between PEP and organizational identification was stronger among employees with high POS than with low POS. Thus, H7 was supported. To examine the PEP's conditional indirect effects on deep acting and surface acting via organizational identification, bootstrapping procedures based on 5000 bootstrapped resamples were used to estimate 95% confidence intervals (CI) for PEP-OI-DA and PEP-OI-SA at high and low POS. As shown in Table 7, under the condition of low POS, PEP's indirect effects on deep acting (indirect effect = 0.091, CI = 0.033, 0.179) and surface acting (indirect effect = −0.090, CI = −0.184, −0.025) via organizational identification were significant. Under the condition of high POS, PEP's indirect effects on deep acting (indirect effect = 0.138, CI = 0.071, 0.205) and surface acting (indirect effect = −0.136, CI = −0.210, −0.054) via organizational identification were significant as well.

Table 6. The moderating effect of perceived organizational support.

Independent Variable		M-7 OI	M-8 OI	M-9 IMM	M-10 IMM
Control variable	Gender	0.008	−0.005	0.012	0.036
	Age	0.126	0.114	−0.153	−0.132
	Tenure	0.002	0.002	0.085	0.084
	Level 1	0.009	0.000	0.229	0.211
	Level 2	−0.053	−0.040	0.016	−0.008
	Level 3	−0.074	−0.081	−0.078	−0.064
PEP		0.577 ***	0.563 ***	0.334 ***	0.361 ***
POS		0.201 ***	0.221 ***	0.005	−0.038
PEP × POS		–	0.083 *	–	−0.154 ***
$R^2/\Delta R^2$		0.308 ***/–	0.316 ***/0.008 **	0.116 ***/–	0.140 ***/0.024 ***

Notes: PEP = perceived external prestige; OI = organization identification; IMM = impression management motive; POS = perceived organizational support; * $p < 0.05$; ** $p < 0.01$; *** $p < 0.001$.

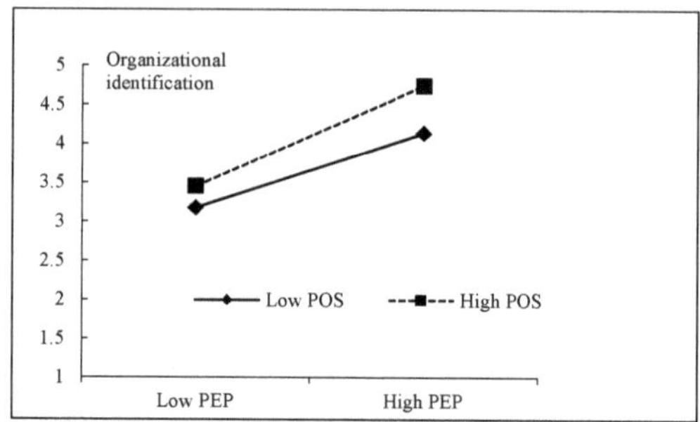

Figure 2. The moderating effect of POS on the relationship between PEP and organizational identification.

Table 7. Results for conditional indirect effects.

Mediators		Indirect Effect	Boot SE	Boot LLCI	Boot ULCI
PEP→OI→DA	POS-1SD	0.091	0.038	0.033	0.179
	POS+1SD	0.138	0.034	0.071	0.205
PEP→OI→SA	POS-1SD	−0.090	0.041	−0.184	−0.025
	POS+1SD	−0.136	0.040	−0.210	−0.054
PEP→IMM→SA	POS-1SD	0.171	0.032	0.112	0.237
	POS+1SD	0.044	0.028	−0.003	0.105

Notes: PEP = perceived external prestige; OI = organization identification; IMM = impression management motive; DA = deep acting; SA = surface acting; POS = perceived organizational support.

Then, we tested the moderating effect of POS on the relationship between PEP and impression management motive. As shown in Table 6, the main effect of POS on impression management motive was not significant ($\beta = -0.038$, $p > 0.05$) and the interaction of PEP and POS on impression management motive was negative and significant ($\beta = -0.154$, $p < 0.001$). The R^2 change between M-9 and M-10 ($\Delta R^2 = 0.024$, $p < 0.001$) was significant, supporting H8. We graphed the simple slopes of PEP on impression management motive at high and low levels of POS. Among frontline employees with low POS, the negative relationship between employees' PEP and impression management motive was stronger than those with high POS (see Figure 3). Because PEP's indirect effect on deep acting via impression management motive is not significant, we did not report PEP's conditional indirect effects on deep acting. Using bootstrapping procedures, PEP's conditional indirect effects on surface acting via impression management motive were examined. As shown in Table 7, under the condition of low POS, PEP's indirect effects on surface acting (indirect effect = 0.171, CI = 0.112, 0.237) was significant. Under the condition of high POS, PEP's indirect effect on surface acting (indirect effect = 0.044, CI = −0.003, 0.105) was not significant.

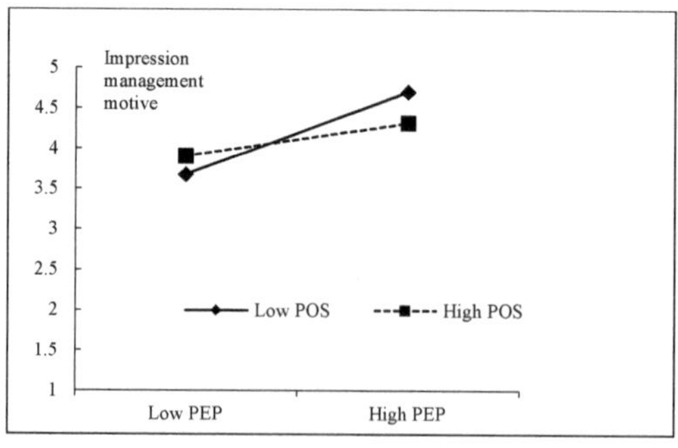

Figure 3. The moderating effect of POS on the relationship between PEP and impression management motive.

5. Discussion

5.1. Theoretical Contributions

Exploring the mechanisms by which perceived external prestige (PEP) influences frontline employees' emotional labor is the purpose of this research. Drawing upon social identity theory and impression management theory, we propose that PEP affects emotional labor via organizational identification and impression management motive. Furthermore, the boundary condition of these mediation effects is examined by introducing perceived organizational support (POS) as a moderator. We contribute to the literature by explaining how and why PEP influence employees' emotional labor.

In particular, our research contributes to the literature on external prestige and emotional labor. First, we contribute to the understanding of the underlying mechanisms by which PEP influences frontline employees' behaviors. Although researchers have argued that PEP influences employees' attitudes and behaviors primarily through organizational identification [13,14], the role of impression management motive has received little attention. We fill this gap by examining the mediation effects of organizational identification and impression management motive simultaneously. It is important to note that PEP's indirect effects via organizational identification and impression management motive on surface acting are negative and positive, respectively. This inconsistent mediation implies that the mechanisms by which PEP influence surface acting are far more complex than we knew. Either ignoring the mediator or just considering one mediator (organizational identification) will underestimate PEP's effect on surface acting. Therefore, our research provided a better and more accurate understanding of the relationship between PEP and employees' emotional labor.

Second, previous research has paid little attention to the conditions under which PEP is more or less likely to influence employees' attitudes and behaviors. Our research demonstrates that POS strengthens (weakens) the indirect effects of PEP on emotional labor through organizational identification (impression management motive). This contributes to a better understanding of the boundary conditions under which PEP fosters frontline employees' engaging in deep acting rather than surface acting.

Third, this study enriches existing literature on the antecedents of employees' emotional labor. Given the importance of employees' emotional labor in service encounters, exploring antecedents of emotional labor has attracted many researchers' attention. Compared with the amount of previous research mainly focused on employee perspective [5,64,65], relatively less attention is given to organizational factors. For example, organizational culture and leadership influence employees' emotional labor [24,66]. Our findings demonstrate that as a form of intangible asset, external prestige can enhance firms' performance by shaping frontline employees' emotional labor.

Finally, our research contributes to the impression management literature by introducing impression management motive as a mediator in our theoretical model. Research on impression management suggests that emotion-expressing skills such as smiling and eye contact are important impression management tactics [16,67], while few studies have empirically examined the relationship between impression management and emotional labor. The results suggest that impression management can serve as a motivation that can drive employees' emotional labor. In addition, different from previous research which focused on impression management in internal organizational contexts, such as interview and supervisor–subordinate relationships [68,69], our research extends impression management theory into the boundary-spanning context, namely frontline employees' interaction with customers during service encounters.

5.2. Managerial Implications

To survive in fiercer competition, service firms are increasingly paying attention to service experiences. The emotions displayed by frontline employees are a key determinant of customer experience in service interactions. Therefore, how to control or intervene in employees' emotional labor is an important and challenging task for managers. Our findings might shed light on how a service firm can intervene in employees' emotional labor using intrinsic motivations. Improving external prestige may be an effective means to promote employees' deep acting. For example, service firms may improve external prestige by communicating the firm's core values, social responsibility, and excellent financial performance to external stakeholders. However, merely improving external prestige is insufficient. Service firms should also pay attention to employee's organizational identification and impression management motive. Specifically, managers need to enhance employees' organizational identification, to guarantee that external prestige can ultimately drive frontline employees to engage in deep acting rather than surface acting. Service firms should also pay attention to POS. In order to improve POS, both managers and colleagues should take measures to support frontline employees. Specifically, managers can change their leadership style, which can evoke employees' perception of respect and trust in the organization. Service firms should value frontline employees' contributions by rewarding and caring for their well-being.

5.3. Limitations and Future Directions

There are several limitations to this research. First, the sample of this research was collected from the hospitality industry in China. Although the hospitality industry probably is an appropriate setting for our research, collecting data from one industry limits the generalizability of the findings. Thus, future research should explore the effects of PEP on frontline employees' emotional labor in different service industries. Second, this research employed a retrospective questionnaire, which is a common cross-sectional design in empirical works. However, a comparison of the survey data carried out over several years would be meaningful. Therefore, future researchers should attempt to integrate experimental and longitudinal approaches, which are conducive to making stronger causal inferences. Third, the present study merely examined the relationship between PEP and frontline employees' emotional labor, while the other consequences of employees' positive behaviors might also be meaningful, such as job crafting. Finally, this research merely focuses on the moderating effect of POS. Future research should investigate other moderators' roles in the relationship between PEP and impression management motive, such as customer orientation.

6. Conclusions

To examine the mediating effect of organizational identification and impression management motivation on the relationship between PEP and emotional labor, as well as the moderating effect of POS, we conducted a series of mediation model analyses and moderated mediation analyses using Hayes' PROCESS macro. Using survey data from 377 frontline employees in 104 Chinese hotels, the results demonstrate that PEP's effects on deep acting are partially mediated by organizational identification. The influence of PEP on surface acting was partly mediated both by organizational identification and impression management motive. A side note of interest is that the indirect effect of PEP on surface acting through organizational identification is negative, while it is positive via impression management motive. Thus, the total effect of PEP on surface acting was underestimated if the mediating role of organizational identification and impression management motive were not considered. Furthermore, the results also suggest that POS positively moderates the influence of PEP on organizational identification, while it negatively moderates the effect of PEP on impression management motive. In other words, for employees with high POS, PEP is more likely to influence their emotional labor via organizational identification, while for employees with low POS, PEP influences frontline employees' emotional labor primarily through impression management motive.

Author Contributions: Conceptualization, P.C.; data curation, Z.L.; formal analysis, P.C.; funding acquisition, P.C.; investigation, P.C.; methodology, J.J.; project administration, P.C.; resources, P.C.; software, J.J.; supervision, P.C.; validation, Z.L.; visualization, J.J.; writing—original draft, J.J. All authors have read and agreed to the published version of the manuscript.

Int. J. Environ. Res. Public Health **2022**, 19, 10778

Funding: This research was funded by the Soft Science Research Program of the Shaanxi Province of China (NO.2021KRM157) and the Scientific Research Program Funded by the Shaanxi Provincial Education Department (Program No.21JZ041).

Institutional Review Board Statement: Ethical review and approval were waived for this study because our research did not entail clinical trials of human or animals, and the questionnaire was formed to investigate frontline employees' opinions on the motivations which drive them to engage in different emotional labor strategies during service encounters, which did not encompass ethical issues.

Informed Consent Statement: Informed consent was obtained from all subjects involved in the study.

Data Availability Statement: Not applicable.

Conflicts of Interest: The authors declare no conflict of interest.

References

1. Grandey, A.A.; Sayre, G.M. Emotional labor: Regulating emotions for a wage. *Curr. Dir. Psychol. Sci.* **2019**, *28*, 131–137. [CrossRef]
2. Hochschild, A.R. *The Managed Heart: Commercialization of Human Feeling*; University of California Press: Los Angeles, CA, USA, 1983.
3. Grandey, A.A. Emotion regulation in the workplace: A new way to conceptualize emotional labor. *J. Occup. Health Psychol.* **2000**, *5*, 85–110. [CrossRef]
4. Liu, X.-Y.; Chi, N.-W.; Gremler, D.D. Emotion Cycles in Services: Emotional Contagion and Emotional Labor Effects. *J. Serv. Res.* **2019**, *22*, 285–300. [CrossRef]
5. Wen, J.; Huang, S.S.; Hou, P. Emotional intelligence, emotional labor, perceived organizational support, and job satisfaction: A moderated mediation model. *Int. J. Hosp. Manag.* **2019**, *81*, 120–130. [CrossRef]
6. Gong, T.; Park, J.; Hyun, H. Customer response toward employees' emotional labor in service industry settings. *J. Retail. Consum. Serv.* **2020**, *52*, 101899. [CrossRef]
7. Kim, J.S. Emotional labor strategies, stress, and burnout among hospital nurses: A path analysis. *J. Nurs. Scholarsh.* **2020**, *52*, 105–112. [CrossRef] [PubMed]
8. Ashtar, S.; Yom-Tov, G.B.; Akiva, N.; Rafaeli, A. When do service employees smile? Response-dependent emotion regulation in emotional labor. *J. Organ. Behav.* **2021**, *42*, 1202–1227. [CrossRef]
9. Medler-Liraz, H.; Seger-Guttmann, T. The joint effect of flirting and emotional labor on customer service-related outcomes. *J. Retail. Consum. Serv.* **2021**, *60*, 102497. [CrossRef]
10. Grandey, A.A.; Fisk, G.M.; Mattila, A.S.; Jansen, K.J.; Sideman, L.A. Is "service with a smile" enough? Authenticity of positive displays during service encounters. *Organ. Behav. Hum. Decis. Processes* **2005**, *96*, 38–55. [CrossRef]
11. Hur, W.-M.; Shin, Y.; Moon, T.W. Linking motivation, emotional labor, and service performance from a self-determination perspective. *J. Serv. Res.* **2022**, *25*, 227–241. [CrossRef]
12. Smidts, A.; Pruyn, A.T.H.; Van Riel, C.B. The impact of employee communication and perceived external prestige on organizational identification. *Acad. Manag. J.* **2001**, *44*, 1051–1062. [CrossRef]
13. Carmeli, A.; Gilat, G.; Weisberg, J. Perceived External Prestige, Organizational Identification and Affective Commitment: A Stakeholder Approach. *Corp. Reput. Rev.* **2006**, *9*, 92–104. [CrossRef]
14. Mignonac, K.; Herrbach, O.; Serrano Archimi, C.; Manville, C. Navigating Ambivalence: Perceived Organizational Prestige–Support Discrepancy and Its Relation to Employee Cynicism and Silence. *J. Manag. Stud.* **2018**, *55*, 837–872. [CrossRef]
15. Mishra, S.K.; Bhatnagar, D.; D'Cruz, P.; Noronha, E. Linkage between perceived external prestige and emotional labor: Mediation effect of organizational identification among pharmaceutical representatives in India. *J. World Bus.* **2012**, *47*, 204–212. [CrossRef]
16. Bolino, M.; Long, D.; Turnley, W. Impression management in organizations: Critical questions, answers, and areas for future research. *Annu. Rev. Organ. Psychol. Organ. Behav.* **2016**, *3*, 377–406. [CrossRef]
17. Wingate, T.G.; Lee, C.S.; Bourdage, J.S. Who Helps and Why? Contextualizing Organizational Citizenship Behavior. *Can. J. Behav. Sci.* **2019**, *51*, 147–158. [CrossRef]
18. Johnson, G.; Griffith, J.A.; Buckley, M.R. A new model of impression management: Emotions in the 'black box' of organizational persuasion. *J. Occup. Organ. Psychol.* **2016**, *89*, 111–140. [CrossRef]
19. Shen, J.; Benson, J. When CSR Is a Social Norm: How Socially Responsible Human Resource Management Affects Employee Work Behavior. *J. Manag.* **2016**, *42*, 1723–1746. [CrossRef]
20. Eisenberger, R.; Rhoades Shanock, L.; Wen, X. Perceived organizational support: Why caring about employees counts. *Annu. Rev. Organ. Psychol. Organ. Behav.* **2020**, *7*, 101–124. [CrossRef]
21. Ashforth, B.E.; Harrison, S.H.; Corley, K.G. Identification in Organizations: An Examination of Four Fundamental Questions. *J. Manag.* **2008**, *34*, 325–374. [CrossRef]
22. Shapoval, V. Organizational injustice and emotional labor of hotel front-line employees. *Int. J. Hosp. Manag.* **2019**, *78*, 112–121. [CrossRef]

23. Nguyen, N.; Besson, T.; Stinglhamber, F. Emotional labor: The role of organizational dehumanization. *J. Occup. Health Psychol.* **2022**, *27*, 179–194. [CrossRef] [PubMed]

24. Lu, J.; Zhang, Z.; Jia, M. Does Servant Leadership Affect Employees' Emotional Labor? A Social Information-Processing Perspective. *J. Bus. Ethics* **2019**, *159*, 507–518. [CrossRef]

25. Carmeli, A. Perceived external prestige, affective commitment, and citizenship behaviors. *Organ. Stud.* **2005**, *26*, 443–464. [CrossRef]

26. Teece, D.J.; Pisano, G.; Shuen, A. Dynamic Capabilities and Strategic Management. *Strateg. Manag. J.* **1997**, *18*, 509–533. [CrossRef]

27. Klimchak, M.; Ward, A.-K.; Matthews, M.; Robbins, K.; Zhang, H. When does what other people think matter? The influence of age on the motivators of organizational identification. *J. Bus. Psychol.* **2019**, *34*, 879–891. [CrossRef]

28. Elstak, M.N.; Bhatt, M.; Van Riel, C.B.; Pratt, M.G.; Berens, G.A. Organizational identification during a merger: The role of self-enhancement and uncertainty reduction motives during a major organizational change. *J. Manag. Stud.* **2015**, *52*, 32–62. [CrossRef]

29. Hogg, M.A.; Terry, D.I. Social identity and self-categorization processes in organizational contexts. *Acad. Manag. Rev.* **2000**, *25*, 121–140. [CrossRef]

30. Ashforth, B.E.; Mael, F. Social identity theory and the organization. *Acad. Manag. Rev.* **1989**, *14*, 20–39. [CrossRef]

31. Cheema, S.; Afsar, B.; Javed, F. Employees' corporate social responsibility perceptions and organizational citizenship behaviors for the environment: The mediating roles of organizational identification and environmental orientation fit. *Corp. Soc. Responsib. Environ. Manag.* **2020**, *27*, 9–21. [CrossRef]

32. Armeli, S.; Eisenberger, R.; Fasolo, P.; Lynch, P. Perceived organizational support and police performance: The moderating influence of socioemotional needs. *J. Appl. Psychol.* **1998**, *83*, 288–297. [CrossRef] [PubMed]

33. Teng, C.-C.; Lu, A.C.C.; Huang, Z.-Y.; Fang, C.-H. Ethical work climate, organizational identification, leader-member-exchange (LMX) and organizational citizenship behavior (OCB): A study of three star hotels in Taiwan. *Int. J. Contemp. Hosp. Manag.* **2020**, *32*, 212–229. [CrossRef]

34. Homburg, C.; Wieseke, J.; Hoyer, W.D. Social Identity and the Service–Profit Chain. *J. Mark.* **2009**, *73*, 38–54. [CrossRef]

35. Maneotis, S.M.; Grandey, A.A.; Krauss, A.D. Understanding the "Why" as Well as the "How": Service Performance is a Function of Prosocial Motives and Emotional Labor. *Hum. Perform.* **2014**, *27*, 80–97. [CrossRef]

36. Bolino, M.C. Citizenship and impression management: Good soldiers or good actors? *Acad. Manag. Rev.* **1999**, *24*, 82–98. [CrossRef]

37. Bozeman, D.P.; Kacmar, K.M. A Cybernetic Model of Impression Management Processes in Organizations. *Organ. Behav. Hum. Decis. Processes* **1997**, *69*, 9–30. [CrossRef]

38. Rosenfeld, P. Impression Management, Fairness, and the Employment Interview. *J. Bus. Ethics* **1997**, *16*, 801–808. [CrossRef]

39. Jones, E.E.; Pittman, T.S. Toward a General Theory of Strategic Self-Presentation. In *Psychological Perspectives on the Self*; Lawrence Erlbaum Associates: Mahwah, NJ, USA, 1982; Volume 1.

40. García-Sánchez, I.-M.; Suárez-Fernández, O.; Martínez-Ferrero, J. Female directors and impression management in sustainability reporting. *Int. Bus. Rev.* **2019**, *28*, 359–374. [CrossRef]

41. Bolino, M.C.; Kacmar, K.M.; Turnley, W.H.; Gilstrap, J.B. A multi-level review of impression management motives and behaviors. *J. Manag.* **2008**, *34*, 1080–1109. [CrossRef]

42. Huffman, B.M.; Hafferty, F.W.; Bhagra, A.; Leasure, E.L.; Santivasi, W.L.; Sawatsky, A.P. Resident impression management within feedback conversations: A qualitative study. *Med. Educ.* **2021**, *55*, 266–274. [CrossRef]

43. Yun, S.; Takeuchi, R.; Liu, W. Employee self-enhancement motives and job performance behaviors: Investigating the moderating effects of employee role ambiguity and managerial perceptions of employee commitment. *J. Appl. Psychol.* **2007**, *92*, 745–756. [CrossRef] [PubMed]

44. Ashforth, B.E.; Humphrey, R.H. Emotional Labor in Service Roles: The Influence of Identity. *Acad. Manag. Rev.* **1993**, *18*, 88–115. [CrossRef]

45. Albrecht, C.-M.; Hattula, S.; Bornemann, T.; Hoyer, W.D. Customer response to interactional service experience: The role of interaction environment. *J. Serv. Manag.* **2016**, *27*, 704–729. [CrossRef]

46. Shumski Thomas, J.; Olien, J.L.; Allen, J.A.; Rogelberg, S.G.; Kello, J.E. Faking It for the Higher-Ups: Status and Surface Acting in Workplace Meetings. *Group Organ. Manag.* **2018**, *43*, 72–100. [CrossRef]

47. Blader, S.L.; Yu, S. Are status and respect different or two sides of the same coin? *Acad. Manag. Ann.* **2017**, *11*, 800–824. [CrossRef]

48. Rhoades, L.; Eisenberger, R. Perceived organizational support: A review of the literature. *J. Appl. Psychol.* **2002**, *87*, 698–714. [CrossRef]

49. Eisenberger, R.; Huntington, R.; Hutchison, S.; Sowa, D. Perceived organizational support. *J. Appl. Psychol.* **1986**, *71*, 500–507. [CrossRef]

50. Kurtessis, J.N.; Eisenberger, R.; Ford, M.T.; Buffardi, L.C.; Stewart, K.A.; Adis, C.S. Perceived organizational support: A meta-analytic evaluation of organizational support theory. *J. Manag.* **2017**, *43*, 1854–1884. [CrossRef]

51. Dutton, J.E.; Dukerich, J.M.; Harquail, C.V. Organizational images and member identification. *Adm. Sci. Q.* **1994**, *39*, 239–263. [CrossRef]

52. Lievens, F.; Slaughter, J.E. Employer Image and Employer Branding: What We Know and What We Need to Know. *Annu. Rev. Organ. Psychol. Organ. Behav.* **2016**, *3*, 407–440. [CrossRef]

53. Crane, A.; Glozer, S. Researching CSR Communication: Themes, Opportunities and Challenges. *J. Manag. Stud.* **2016**, *53*, 1223–1252. [CrossRef]
54. Zavyalova, A.; Pfarrer, M.D.; Reger, R.K. Celebrity and Infamy? The Consequences of Media Narratives about Organizational Identity. *Acad. Manag. Rev.* **2017**, *42*, 461–480. [CrossRef]
55. Brotheridge, C.M.; Lee, R.T. Development and validation of the Emotional Labour Scale. *J. Occup. Organ. Psychol.* **2003**, *76*, 365–379. [CrossRef]
56. Mael, F.; Ashforth, B.E. Alumni and their alma mater: A partial test of the reformulated model of organizational identification. *J. Organ. Behav.* **1992**, *13*, 103–123. [CrossRef]
57. Herrbach, O.; Mignonac, K.; Gatignon, A.-L. Exploring the role of perceived external prestige in managers' turnover intentions. *Int. J. Hum. Resour. Manag.* **2004**, *15*, 1390–1407. [CrossRef]
58. Eisenberger, R.; Armeli, S.; Rexwinkel, B.; Lynch, P.D.; Rhoades, L. Reciprocation of Perceived Organizational Support. *J. Appl. Psychol.* **2001**, *86*, 42–51. [CrossRef]
59. Rich, B.L.; Lepine, J.A.; Crawford, E.R. Job engagement: Antecedents and effects on job performance. *Acad. Manag. J.* **2010**, *53*, 617–635. [CrossRef]
60. Jung, Y.; Takeuchi, N. A lifespan perspective for understanding career self-management and satisfaction: The role of developmental human resource practices and organizational support. *Hum. Relat.* **2018**, *71*, 73–102. [CrossRef]
61. Podsakoff, P.M.; Organ, D.W. Self-reports in organizational research: Problems and prospects. *J. Manag.* **1986**, *12*, 531–544. [CrossRef]
62. Hayes, A.F. *Introduction to Mediation, Moderation, and Conditional Process Analysis: A Regression-Based Approach*; Guilford Publications: New York, NY, USA, 2013.
63. Aiken, L.S.; West, S.G.; Reno, R.R. *Multiple Regression: Testing and Interpreting Interactions*; Sage: Thousand Oaks, CA, USA, 1991.
64. Hwang, P.-C.; Han, M.-C. Does psychological capital make employees more fit to smile? The moderating role of customer-caused stressors in view of JD-R theory. *Int. J. Hosp. Manag.* **2019**, *77*, 396–404. [CrossRef]
65. Doucet, L.; Shao, B.; Wang, L.; Oldham, G.R. I know how you feel, but it does not always help: Integrating emotion recognition, agreeableness, and cognitive ability in a compensatory model of service performance. *J. Serv. Manag.* **2016**, *27*, 320–338. [CrossRef]
66. Barsade, S.G.; O'Neill, O.A. What's Love Got to Do with It? A Longitudinal Study of the Culture of Companionate Love and Employee and Client Outcomes in a Long-term Care Setting. *Adm. Sci. Q.* **2014**, *59*, 551–598. [CrossRef]
67. Stevens, C.K.; Kristof, A.L. Making the right impression: A field study of applicant impression management during job interviews. *J. Appl. Psychol.* **1995**, *80*, 587–606. [CrossRef]
68. Bourdage, J.S.; Schmidt, J.; Wiltshire, J.; Nguyen, B.; Lee, K. Personality, interview performance, and the mediating role of impression management. *J. Occup. Organ. Psychol.* **2020**, *93*, 556–577. [CrossRef]
69. Chen, Z.; Huo, Y.; Lam, W.; Luk, R.C.T.; Qureshi, I. How perceptions of others' work and impression management motives affect leader–member exchange development: A six-wave latent change score model. *J. Occup. Organ. Psychol.* **2021**, *94*, 645–671. [CrossRef]

International Journal of
Environmental Research and Public Health

Article

Ergonomic Factors That Impact Job Satisfaction and Occupational Health during the SARS-CoV-2 Pandemic Based on a Structural Equation Model: A Cross-Sectional Exploratory Analysis of University Workers

Víctor Manuel Ramos-García [1], Josué Aarón López-Leyva [2,*], Raúl Ignacio Ramos-García [3], Juan José García-Ochoa [1], Iván Ochoa-Vázquez [1], Paulina Guerrero-Ortega [4], Rafael Verdugo-Miranda [1] and Saúl Verdugo-Miranda [1]

1   Departamento de Física, Matemáticas e Ingeniería, Universidad de Sonora, Navojoa 85880, Mexico
2   Centro de Innovación y Diseño, CETyS Universidad, Ensenada 22860, Mexico
3   Independent Researcher, Long Beach, CA 90813, USA
4   Academia de Matemáticas, Centro de Bachillerato Tecnológico Industrial y de Servicios No. 64, Navojoa 85860, Mexico
*   Correspondence: josue.lopez@cetys.mx

Citation: Ramos-García, V.M.; López-Leyva, J.A.; Ramos-García, R.I.; García-Ochoa, J.J.; Ochoa-Vázquez, I.; Guerrero-Ortega, P.; Verdugo-Miranda, R.; Verdugo-Miranda, S. Ergonomic Factors That Impact Job Satisfaction and Occupational Health during the SARS-CoV-2 Pandemic Based on a Structural Equation Model: A Cross-Sectional Exploratory Analysis of University Workers. *Int. J. Environ. Res. Public Health* **2022**, *19*, 10714. https://doi.org/10.3390/ijerph191710714

Academic Editors: Masahito Fushimi and Paul B. Tchounwou

Received: 3 August 2022
Accepted: 24 August 2022
Published: 28 August 2022

Publisher's Note: MDPI stays neutral with regard to jurisdictional claims in published maps and institutional affiliations.

Abstract: This paper presents a structural equation model to determine the job satisfaction and occupational health impacts concerning organizational and physical ergonomics, using (as a study) objective unionized workers from the University of Sonora, South Campus, as an educational enterprise, during the SARS-CoV-2 pandemic. The above is a key element of an organizational sustainability framework. In fact, there exists a knowledge gap about the relationship between diverse ergonomic factors, job satisfaction, and occupational health, in the educational institution's context. The method used was a stratified sample of workers to which a job satisfaction–occupational health questionnaire was applied, consisting of 31 items with three-dimensional variables. As a result, the overall Cronbach's Alpha coefficient was determined, 0.9028, which is considered adequate to guarantee reliability (i.e., very high magnitude). Therefore, after the structural equation model, only 12 items presented a strong correlation, with a good model fit of 0.036 based on the root mean square error of approximation, 1.09 degrees of freedom for the chi-square, 0.9 for the goodness of fit index, and a confidence level of 95%. Organizational and physical factors have positive impacts on job satisfaction with factor loads of 0.37 and 0.53, respectively, and *p*-values of 0.016 and 0.000, respectively. The constructs related to occupational health that are considered less important by the workers were also determined, which would imply a mitigation strategy. The results contribute to the body of knowledge concerning the ergonomic dimensions mentioned and support organizational sustainability improvements in educational institutions and other sectors.

Keywords: structural equation model; job satisfaction; organizational ergonomics; physical ergonomics

1. Introduction

Since the 1930s, there has been great interest in job satisfaction research, which probably peaked in the 1960s. In the 1980s, this issue began to be more closely related to the quality of life at work, its impact on mental health, and the relationships between coworkers and family, with a growing concern for the individual's personal development in an educational context throughout life [1]. Nowadays, job satisfaction is one of the most important issues according to the subjective perception of the worker. Therefore, researchers have focused on developing studies of this nature. In fact, human resources are of great value within organizations; the impact can be measured through the business transformation, where the success of the company is guaranteed through experience, knowledge, motivation, and the management of labor environment changes [2]. Hence, it makes it possible for organizations to achieve their business objectives and results through their employees, based on the relevance of seeing workers as valuable capital, which determines an important weight in the achievement of the objectives to follow [3]. Thus, it has been found that job satisfaction is an important and highly beneficial element for both organizations and employees in an organizational sustainability framework [4,5]. Generally, this element is a pleasant or positive emotional state of subjective perception, resulting from the evaluation of work experiences, and it heavily impacts the behavior of the worker [4,6,7]. In particular, organizational sustainability can be understood as the organization's capacity based on its internal and external processes in order to maintain and improve its operation mode to increase performance, profitability, and competitiveness.

Around the world, interest in the work environment in institutions/organizations has gained enormous relevance due to the various problems faced, largely due to internal problems [8]. Inside them, one of the main issues is the lack of job satisfaction, which inhibits the development of creative and innovative work [9]. In consequence, job dissatisfaction affects the worker's performance as well as productivity, causing demotivation and a lack of interest in his/her work, which produces worker apathy and possibly implies that the worker is not correctly complying with his/her daily or habitual assigned functions [10]. On the other hand, this situation can lead to anxiety or stress and, in extreme cases, the worker can suffer from depression [11]. Universities (or academic institutions) are no exception, although it is true that they generate professionals in specialized areas, such as manufacturing, engineering, and services, among others, they must also improve their internal production processes in the workplace as part of an organizational sustainability framework, supporting the regional sustainability development [12].

Furthermore, the factors that influence job satisfaction are essential to improve the well-being of (a large part of) society and high-performance jobs [5,13]. At the same time, a satisfaction index can be established based on the working conditions, allowing one to determine the main deficient elements in which action must be taken to achieve improvements in the work environment. In addition, these factors constitute great important elements for the development of all processes where human resources intercede. Therefore, the measurement of these elements is important to determine the relationships of the factors that most affect and impact the job satisfaction of the study object. The literature shows that the study of job satisfaction has been approached from multiple dimensions, and is linked to some variables, such as motivation, sociocultural aspects, and economic and communication features, among others [10]. However, it has not been addressed with ergonomic approaches in higher-level institutions focused on staff who are unionized and provide operational services (maintenance, security, infrastructure hygiene, drives, and others) because this scenario is not widely studied, i.e., case studies and related information are scarce [14,15]. Therefore, this study aims to determine the impacts of ergonomic factors on job satisfaction and occupational health concerning unionized workers who provide operational services at an institution through structural equation analysis.

The rest of this manuscript is organized as follows: In Section 2, the contextual and theoretical backgrounds are explained. In particular, Section 2.1 describes the dimensions of ergonomics. Section 2.2 describe the structural equation model, Section 2.3 explain the exploratory factor analysis, and Section 2.4 describe the confirmatory factor analysis, as the mathematical methods that will be used. Section 3 presents the materials and methods applied. Section 4 presents the instrument validation. Section 5 presents a brief discussion of the results and implications. Finally, the conclusions and recommendations are presented in Section 6.

2. Theoretical Background

2.1. Dimensions of Ergonomics and Hypotheses Development

Ergonomics has been universally used to improve the quality of human life, such as health, safety, comfort, and productivity so that the personnel is satisfied in the environments and work activities. In general, ergonomics can be analyzed considering two approaches (i.e., dimensions), the organizational dimension (organizational ergonomics, OE) and the physical dimension (physical ergonomics, PE), which are highly related to organizational sustainability. Organizational ergonomics, also called macroergonomics, refers to the optimization of social–technical systems, including their organizational structures, policies, and processes [16–18]. The relevant issues include communications, management of resources, labor projects, temporal work organization, teamwork, participative project, new work paradigms, cooperative work, organizational culture, network organizations, and quality management. The macroergonomics approach satisfies the correct criteria for the design of work systems and work designs with the man–system interface; focused on the man, applying a "humanized" task to the role of different assignments [19]. Therefore, job designs also include work modules, tasks, knowledge, and skill requirements, as well as factors such as an autonomy degree, identity, feedback, and opportunities for social interactions. The purpose of organizational ergonomics involves the optimization of the design of social–technical work systems and the study of the effect of organizational structures on human behavior and safety. This goal has been achieved through the systematic consideration of the relevant variables of sociotechnical systems in the ergonomic analysis, design, implementation, evaluation, and control of the process. These variables correspond to the technological subsystem, personal subsystem, and external environment [20]. To achieve the maximum objectives of ergonomics (i.e., optimize the well-being of people and the overall performance of the system) the interaction between the subsystems has to be functional, respecting the capacities and limitations of the human being and his/her culture [21]. Considering the above, managers need to recognize that job satisfaction is a feeling of relative pleasure or pain [1]. Furthermore, this factor arises from the perception of how an employee views his/her job, positively or negatively, or how he/she likes or dislikes the job as a result of the employee's perception of the job. To summarize, satisfaction is considered one of the most important pillars in assessing the success of companies and it can help management understand the reactions of workers to their jobs. Hence, the hypothesis related to job satisfaction and organizational ergonomics is presented:

Hypothesis 1 (H1). *The ergonomic factors inherent to the **process** have positive impacts on the job satisfaction of workers who provide operational services in the institution.*

Regarding physical ergonomics, it is considered of interest for the anthropometry, anatomy, and biomechanics of man, concerning physical, mental, and environmental efforts. Moreover, it is a multidisciplinary model that is interested in the adaptation of work for man and how these relate to the physical activities involved in the use of the musculoskeletal and cardiovascular systems [22,23]. This construct was established as the science that studies the dimensions of the human body, the knowledge, and techniques to carry out measurements, as well as their statistical treatments [24]. Thus, it seeks to provide anthropometric data that serve as the bases for sizing objects that adjust to the true characteristics of end users. In addition, considering the reality of various industrial sectors, most work tasks require the worker to maintain a fixed posture for long periods, and if a poorly designed position is added to this, then it does not correspond to the anthropometric characteristics of the end users; consequently, it can encourage the adoption of uncomfortable postures, undue efforts, causing discomfort, fatigue in certain muscle groups, and health effects on workers [25]. Moreover, productivity as well as quality decrease. Finally, the probability of errors and the number of work accidents would increase. This field perceives the handling of loads, repetitive movements, and inadequate postures, which cause musculoskeletal difficulties and pathological alterations in health [22]. This construct has a considerable weight in satisfaction. Therefore, the forms that job satisfaction take in objective conditions are related to occupational safety and hygiene, workloads, occupational health, etc., and in subjective conditions of the worker in the sense of how they experience it. The occupational risks of physical loads should be included in the measurement of job satisfaction since it has been shown that a dissatisfied worker is more likely to suffer accidents. Ergonomics at work becomes relevant when identifying a situation that causes a deterioration in the employee's health. Therefore, the following hypothesis is presented:

Hypothesis 2 (H2). *The ergonomic factors inherent to the **operator** have positive impacts on the job satisfaction of the workers who provide operational services in the institution.*

In this research, the third hypothesis relates the two constructs of organizational and physical ergonomics—to determine if there is a relationship in both latent variables. Thus, the third hypothesis is:

Hypothesis 3 (H3). *The ergonomic factors inherent to the **process** (related to OE) have positive impacts on the ergonomic factors inherent to the **operator** (related to PE) who provides operational services at the institution.*

Figure 1 shows the relationship between the mentioned hypotheses.

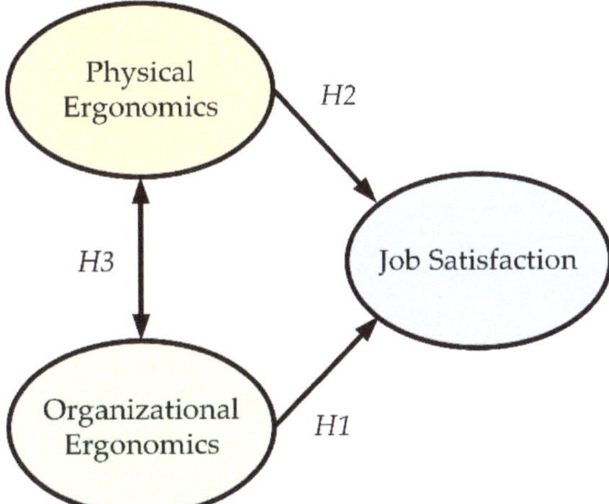

Figure 1. Relationship between ergonomic aspects and job satisfaction.

2.2. Structural Equation Model

The structural equation model (SEM) is a multivariate technique that is applied in research in various disciplines due to its ability to explain causal relationships between qualitative and quantitative variables to test theoretical models [26]. The main contribution of SEM is that it allows researchers to evaluate theoretical models; it is becoming one of the most powerful tools for the study of causal relationships on non-experimental data when these relationships are linear [27]. This model designates a set of procedures and techniques for multivariate statistics that include a large number of classical methods, such as linear regression, exploratory and confirmatory factor analysis, and path analysis, among others [28]. However, its main feature involves the possibility of including unobservable variables in a much wider variety of models. Thus, it is a clear and objective tool for the empirical testing of theoretical hypotheses, being used in several disciplines, such as economics, psychology, sociology, education, marketing, etc. [29–32]. In this research, as part of SEM, we use (as a first step) the exploratory factor analysis and, consequently, the confirmatory factor analysis to eliminate those variables that do not have strong correlations in the study and determine which items of the ergonomic aspects have more relations and impacts with job satisfaction.

2.3. Exploratory Factor Analysis

To explore the latent variables more precisely, we use the exploratory factor analysis (EFA), which is one of the most frequently applied techniques in studies related to the development and validation of different areas since it explores the set of latent variables or common factors that explain the responses of the items, which are revealed from the observed variables [33,34]. In particular, the objective of this technique was to analyze and validate factors underlying a large set of data and reduce a large number of operational indicators into smaller numbers of conceptual variables [35]. Therefore, we chose—as a criterion—to accept those items whose values are greater than or equal to 0.5 [36]. Moreover, this multivariate method allows for group variables (e.g., items) that are strongly correlated with each other, and whose correlations with the variables of other groups (factors) are lower. Although the variables used are generally continuous, it is also possible to use this method on categorical variables [37]. According to [38], through the EFA, the score variability of a set of variables is explained by a smaller number of dimensions or factors. In this way, for example, a large number of items can be reduced to a small number of factors or dimensions that confer a theoretical meaning to the measurement. Each of these factors can group the intercorrelated items that are, at the same time, relatively independent from others sets (factors) of items. In addition, another factor that currently affects job satisfaction (in all sectors of society and industry) is related to the SARS-CoV-2 pandemic [39–41]. In particular, several workplaces have analyzed this aspect. However, none are based on the exploratory factor analysis.

2.4. Confirmatory Factor Analysis

The confirmatory factor analysis (CFA) allows correcting or corroborating if there is a deficiency of the EFA, leading to further testing of the specified hypotheses [42]. It also analyzes the covariance matrix instead of the correlation matrix, which helps to establish whether the indicators are equivalent [43]. In particular, The CFA is represented by flow diagrams (path diagrams), according to its particular specifications (see Figure 2). The rectangles represent the items and the ellipses, the common factors. The unidirectional arrows between common factors and items express saturations and the bidirectional arrows indicate the correlation between common or unique factors [44]. Thus, this method provides the statistical framework to evaluate the validity and reliability of each item instead of performing a global analysis, helping the researcher to optimize both the construction of a measurement instrument and the analysis of results [45]. The mathematical equation of the model is described by $n = \lambda_i \xi_i + \delta_i$, where n is the number of all the observable variables that appear in each dimension. In this study, 31 items are presented as observable variables, λ_i is the factor load that explains the item concerning the latent variable, ξ_i is the latent variable, in which this research has three dimensions (job satisfaction related to physical and organizational ergonomics), and finally δ_i is the error of the observable variable. Moreover, it is important to mention that no studies were found that relate ergonomic factors in the three dimensions mentioned in the structural equation model framework. Furthermore, there are no works related to job satisfaction research educational institutions using this model during the SARS-CoV-2 pandemic. Considering the above, particularly the lack of projects related to institutions in the educational sector, the purpose of this study was to identify which factors have strong correlations to determine the impact of job satisfaction (JS) concerning ergonomic factors (organizational and physical dimensions), using (as study objects) the unionized workers at the University of Sonora, South Campus, through a cross-sectional exploratory factor analysis.

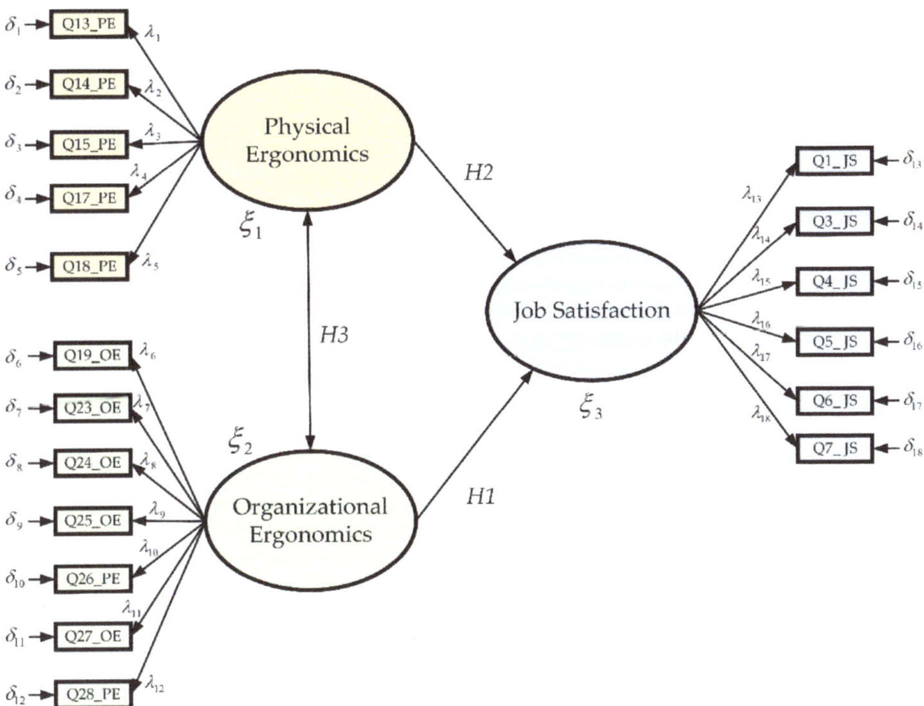

Figure 2. The first test of the model.

3. Materials and Methods

For this work, the study design was a non-experimental quantitative type, cross-sectional, with a correlational scope; the study objects were unionized workers who provide operational manual services at a public higher education school. STATA 14 software (StataCorp LLC, College Station, TX, USA) was used for the descriptive data analysis and the implementation of different estimation techniques. The number of employees integrated is represented by a population of 92 unionized people. Furthermore, the sample size was determined to know the exact number of participants to be included in job satisfaction research, which was divided into groups with different degrees—from the basic level to the undergraduate level using the finite population formula, as Table 1 shows. Thus, stratified sampling was selected because the study object had the same chance of being selected randomly. Consequently, the standard deviation (*sh*) was used in the stratified sample to know the exact number of workers for each category, where *n* is the sample of the study (75 unionized people); it was divided into *N* (the population of 92 unionized people) to have 0.815 (*n/N*). As a result, the number of people by category who answered the questionnaire designed in a structured interview framework for this research is shown in Table 2.

Table 1. Subpopulation samples for educational levels based on the Mexican education system.

Educational Level	Amount of People (*Nh*)
Primary (MX)/Elementary (US)	14
Secondary (MX)/Middle (US)	40
Preparatory (MX)/High School (US)	19
Undergraduate (MX)/College (US)	19
Total	**92**

Table 9. Constructs of organizational ergonomics.

Dimension	Items
Organizational Ergonomic	Q19_OE: The communication with bosses. Q20_OE: The communication with colleagues. Q21_OE: Teamwork. Q22_OE: Your work Schedule. Q23_OE: My roles and responsibilities are well defined. Q24_OE: The new paradigms in my work. Q25_OE: Suggestions and ideas are taken into account in my work team. Q26_OE: Quality improvement initiatives. Q27_OE: The way in which the institution is managed. Q28_OE: Resource management. Q29_OE: In my job, I can develop my skills. Q30_OE: I receive information on how I perform my work.Q31_OE: Medical Services.

The survey also included some demographic variables (gender, antiquity, schooling level, place of birth, functional area, age, type of contract, civil status, place of residence, and salary), since they are factors that conform to the different characteristics of human populations [48].

4. Instrument Validation

The correlation calculation supported the validity verification of the study concerning the construct of the job satisfaction (*JS*) variable to the ergonomic aspects inherent to the operator (physical) and the processes (organizational). Table 10 shows the Pearson correlation and significance level results concerning the construct and the dimensions mentioned. The significance level (p) was typically set no higher than 0.05.

Table 10. Pearson correlation coefficient calculation concerning the constructs based on the sample (75 workers).

Variable	OE	JS	PE
Organizational ergonomic (*OE*)	1	0.707 *	0.552 *
Job satisfaction (*JS*)	0.707 *	1	0.628 *
Physical ergonomic (*PE*)	0.552 *	0.628 *	1

* $p < 0.05$.

Regarding the job satisfaction–organizational ergonomics relationship, it presented a correlation coefficient (Pearson correlation) value of 0.707 (high and significant positive correlation), which means that the more this ergonomic aspect is linked, the more the workers will be satisfied in this institution. Consequently, it has high validity in the relation between these two variables. Moreover, the Pearson correlation between job satisfaction and physical ergonomics is 0.628, which means that the more the conditions for physical workloads are improved, the better the worker´s job satisfaction will be. Likewise, these two variables had a high validation in the investigation. Moreover, the relationship between organizational ergonomics and physical ergonomics showed a moderate positive correlation of 0.552 with a significant level and it was directly proportional to the two variables. Hence, if the perceptions of both variables increase, employee satisfaction will increase.

5. Results and Discussion

Once the results related to the descriptive analyses of the variables were obtained, the next step was to determine the items that belonged to each construct through the component matrix, in order to establish the appropriate instrument. For this, the criterion was taken to accept those items whose values were greater or equal to 0.5, since explanatory capacity was gained.

Table 11 shows the results, taking into account the KMO–Bartlett test, where 13 observable variables were eliminated from the 31 original items, i.e., 31 items were analyzed for EFA. Therefore, the overall KMO–Bartlett test of the instrument was 0.789, with the KMO–Bartlett value of the JS, PE, and OE being 0.744, 0.798, and 0.825, respectively. To clarify, regarding the job satisfaction (JS) variable, out of a total of 10 items, only 6 of these had strong correlation coefficients between Q6_JS, Q3_JS, Q1_JS, Q4_JS, Q7_JS, and Q5_JS. While the items Q2_JS, Q8_JS, Q9_JS y Q10_JS were eliminated because these variables obtained values less than 0.5. Concerning the physical ergonomic (PE) construct, out of about eight items, only five (Q17_PE, P18_PE, Q15_PE, Q14_PE, and Q13_PE) had strong correlation coefficients between the reagents. While the items Q11_PE, Q12_PE y Q16_PE were eliminated using the same elimination rule mentioned. Similarly, for the organizational ergonomic (OE) dimension, seven items (Q24_OE, Q26_OE, Q28_OE, Q19_OE, Q23_OE, Q27_OE, and Q25_OE) had strong correlations, deleting the items, Q19_OE, Q20_OE, Q21_OE, Q22_OE, Q29_OE, and Q30_OE using the same criteria. In the previous procedure, the exploratory factor analysis was developed to know and adjust the items that support the construction of the proposed model, resulting in a total of 18 endogenous variables among the three constructs. Once this part of the study was completed, the next step was to apply the confirmatory factor analysis using the 18 items that were the EFA results. This analysis made it possible to corroborate or correct (if needed) the deficiency of the FEA, leading to further testing of the specified hypotheses [42]. In the CFA, it is necessary to observe the factor loadings that allow the correlation between variables and factors to be established. The closer they are to the unit (1), the higher the correlation. A rule of thumb in the CFA states that loadings must be ≥ 0.3 in the absolute value considered optimal [36,37]. Consequently, Figure 2 shows the first test of the proposed model as a whole of the constructs with the observable variables, of which the physical ergonomics variable had five items; concerning organizational ergonomics—seven items; finally, job satisfaction was represented by six items, for the verification of the hypotheses raised. The following model was developed, taking into account the results of the present research.

For the model validation, the observable variables having loading ≥ 0.3 with the CFA were considered. Table 12 presents the criteria for each absolute fit index to verify whether the study has a good or acceptable fit to the model, which means a *p*-value of less than 0.05.

A detailed analysis of the proposed model was carried out in order to verify which observable variables were relevant in each construct for the validation of the hypotheses put forward. Hence, from a total of 18 variables (considering 31 items as inputs) resulting from the exploratory factor analysis, 6 of them were eliminated in the first test of the confirmatory factor analysis for having low factor loadings of 0.3 and also for not complying with adequate adjustments considering the diverse goodness of fit analysis. These were Q1_JS, Q3_JS, Q5_JS, Q14_EF, Q23_EO, and Q28_EO. In the second test of the confirmatory factor analysis, 12 variables as inputs were considered, and then the goodness of fit and the hypotheses were analyzed to determine whether the model was accepted or rejected. Thus, Figure 3 shows the final test (based on CFA) of the proposed model. To support the study, chi-square was used as a hypothesis test, which compared the observed distribution with an expected distribution of the data, whose purpose was to test the relationship between the two variables. The goodness of fit index (GFI) is an index used to measure and compare the discrepancies between various constructs in a fitted model, and standardized (SRMR) refers to the standardized root mean square residual, which indicates that if the index is closer to 0, the model will be better.

Table 11. Component matrix per construct.

Items	JS	PE	OE
Q6_JS	0.807		
Q3_JS	0.783		
Q1_JS	0.765		
Q4_JS	0.731		
Q7_JS	0.643		
Q5_JS	0.565		
Q17_PE		0.809	
Q18_PE		0.791	
Q15_PE		0.741	
Q14_PE		0.699	
Q13_PE		0.691	
Q24_OE			0.750
Q26_OE			0.747
Q28_OE			0.729
Q19_OE			0.689
Q23_OE			0.671
Q27_OE			0.669
Q25_OE			0.667

Table 12. Absolute fit index.

Absolute Fit Index	Good Fit	Acceptable Fit
Chi-square $(x^2)/df$	$0 \leq x^2 \leq 2df$	$2df \leq x^2 \leq 3df$
Root mean square error of approximation ($RMSEA$)	$0 \leq RMSEA \leq 0.05$	$0.05 \leq RMSEA \leq 0.08$
Goodness of fit index (GFI)	$0.95 \leq GFI \leq 1.00$	$0.90 \leq GFI \leq 0.95$
Standardized ($SRMR$)	$0 \leq SRMR \leq 0.05$	$0.05 \leq SRMR \leq 0.10$

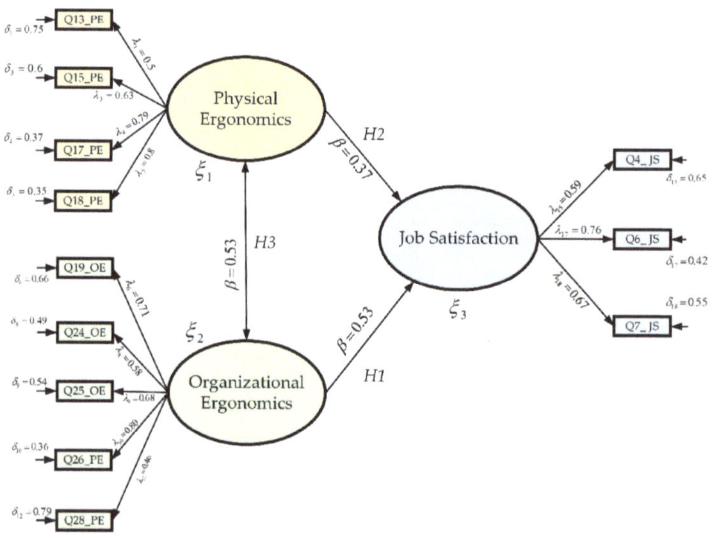

Figure 3. The final test concerning JS with PE and OE.

5.1. Goodness of Fit Results

With respect to the results obtained in the second test of the model, the evaluation of the goodness of fit is that the RMSEA has a value of 0.036, and a chi-square (x^2) of 1.09 degrees of freedom (df), which means that both have a good fit in the model. In addition, the standardized root mean square (SRMR) presents an index of 0.066, and the goodness of fit index (GFI) of about 0.900, which infers that the study has an acceptable fit in the research. Consequently, it presents favorable results concerning the variables proposed in the model. Moreover, the results of the hypotheses show good results in the hypotheses proposed, as Table 13 shows for $p < 0.05$.

Table 13. Model regression coefficients.

Model Regression Coefficients	β	p
Job Satisfaction ← Organizational Ergonomic	0.53	0.000
Job Satisfaction ← Physical Ergonomic	0.37	0.016
Physical Ergonomic ← Organizational Ergonomic	0.53	0.000

5.2. Hypothesis Analysis

In the construct, organizational ergonomics were correlated with five indicators, which were: communication with managers (Q19_OE), new paradigms at work (Q24_OE), suggestions and ideas taken in the work team (Q25_OE), quality improvement initiatives (Q26_OE), and the way the institution was managed (Q27_OE), with regression weights values (λ) of 0.71, 0.58, 0.68, 0.80 and 0.46, respectively. In this dimension, the variables that most explain the organizational ergonomics were related to the items: quality improvement initiatives (related to Q27_OE) and communication with managers (related to Q19_OE). Therefore, to achieve quality improvement, the institution must implement good communication between the employee and the bosses because otherwise this variable would be affected; consequently, there would be a lack of quality improvement. On the other hand, the suggestions and ideas item was used by the work team (related to Q25_OE), in this aspect, the institution should take into account the ideas provided by the workers since they are generally aimed at improving the quality of work within the institution. Finally, there is a significant effect between EO and JS, resulting in a p-value of 0.000. As a result, the EO has a direct positive relationship and impact with JS because the standardized regression coefficient is $\beta = 0.53$. The aforementioned indicates that Hypothesis 1 (*H1*) is accepted.

According to the physical ergonomics latent variable, it is correlated with four indicators, which are: repetitive movements in the workplace (Q13_PE), physical stress, such as strain, neck, shoulder, and back loads (Q15_PE), workloads are well distributed (Q17_PE), and physical demands in the workplace (Q18_PE), where the results in regression weights (λ) are 0.50, 0.63, 0.79 and 0.80, respectively. Considering all the variables in this construct, the items that most explain physical ergonomics are the physical demands in the workplace (related to Q18_PE), and the distribution of the workload (related to Q17_PE). Thus, with respect to the physical demands that are contemplated in the institution, they must be assessed for each employee in the corresponding work area to avoid injuries that could put both the worker and the institution at risk. On the other hand, the institution must take care of the distribution of workloads since they must be equitable for each worker. Therefore, there is a significant effect between PE and JS, given that the p-value is 0.016. This means that physical ergonomics has a direct positive relationship with job satisfaction because the standardized regression coefficient is 0.37 (i.e., $\beta = 0.37$). The aforementioned indicates that Hypothesis 2 (*H2*) is accepted.

Finally, the relationship between the latent variables—organizational ergonomics and physical ergonomics—is significant (i.e., 0.000). Therefore, OE has a direct positive relationship with PE (i.e., $\beta = 0.53$). In addition, it can be mentioned that OE has a positive indirect relationship with JS through PE. Thus, the aforementioned indicates that Hypothesis 3 (*H3*) is accepted.

5.3. Impact on the Organizational Sustainability and Implications

This study has important implications for organizational sustainability policy and practice. In fact, it allows establishing a quantitative tool that helps analyze job satisfaction as a basic element for the development of particular guidelines in the financial, social, and environmental spheres. The foregoing implies that any initiative related to sustainable development that does not have the appropriate impetus on the part of the staff would be destined to fail. However, as many international standards exemplify (e.g., ISO-50001, energy management system), the encouragement and participation of staff are crucial for the success of sustainable development in a leadership framework. In addition, this study makes it possible to significantly impact occupational safety, and it addresses sustainable development from a more human perspective.

Regarding the theoretical and methodological implications, this study contributes to the organizational sustainability literature via a formal mathematical analysis of a particular business sector, in our case, the educational sector. However, this contribution can easily be adapted to other business sectors since the ergonomic dimensions analyzed are inherent to people and organizations. In particular, according to the particular necessities, some improvements in the method and instruments are feasible. Regarding the methodological implications, it is important to mention that within the organizational sustainability framework, the methods that support this development must be continuously enhanced. In this way, this study proposes a formal mathematical analysis that could help the permanent monitoring of the latent variables mentioned as part of the organizational sustainability methodology adopted by each company.

In the same sense, this study also presents very interesting practical implications. For example, the constant monitoring of the mentioned latent variables would allow knowing the job satisfaction level of particular periods and, thus, establish a job satisfaction baseline, which can serve to establish actions as countermeasures to improve job satisfaction level.

Finally, this study presents some limitations that establish potential future research. One of the main aspects is the number of people considered in the study. This implies that it is not possible to increase the organizational sustainability level (considering a baseline) only by analyzing a particular group of personnel; that is, the study must be extended to more workers and from other departments.

5.4. Impact on the Occupational Health

The above results allow us to infer that the work environment during the pandemic affected the job satisfaction of the interviewees. In particular, changes made during the pandemic related to physical and organizational ergonomics that affected job satisfaction. Although a specific study related to cognitive ergonomics was not carried out, which studies the cognitive aspects of workers and the interactions with the work system, qualitatively, it was perceived that job stress levels increased during the pandemic, which is related to the variables that most impact job satisfaction. In the same sense, the educational institution did not carry out a formal deployment of coping strategies against stress in workers, which was reflected in the results of job satisfaction. In addition, the results shown in Table 11 regarding physical ergonomics imply that the constructs (i.e., Q11_PE: Safety in your workplace, Q12_PE: Hygiene in your workplace, Q16_PE: Workplace design and Q18_PE: In general, how satisfied are you with the physical demands in your workplace?) do not have greater relevance for occupational health and job satisfaction. The foregoing imposes important organizational challenges regarding risk analysis. In a similar sense, regarding organizational ergonomics, the constructs that are least perceived as important for job satisfaction and occupational health are L Q20_OE: The communication with colleagues, Q21_OE: Teamwork, Q22_OE: Your work Schedule, Q29_OE: In my job, I can develop my skills, Q30_OE: I receive information on how I perform my work, and Q31_OE: Medical Services. Being the Q31_OE is more surprising.

6. Conclusions

The structural equation model is a useful tool that allows us to identify and group indicators that are strongly correlated with each other to reduce variables that do not contribute significantly to the study. With the development of this research, it was possible to know the parameters that have close relationships with the dimensions already raised during

Int. J. Environ. Res. Public Health **2022**, *19*, 10714

the SARS-CoV-2 pandemic. Nowadays, there is a lack of studies about the dimensions of ergonomics regarding job satisfaction–occupational health and other variables. Thus, our results open the door to developing other multivariate statistical methods as the next step to have a more in-depth analysis of the constructs of job satisfaction concerning other ergonomic aspects, such as cognitive and temporary ergonomics. Moreover, the results presented can be applied as part of the design, planning, and management of technical and social systems at any organization. Thus, this study can be applied to the managers of institutions in order for them to know the perceptions of their workers who provide operational services. Therefore, it would be a matter of interest to know the perceptions or feelings of both parties. Finally, this methodology and finding can be considered as a background to other sectors, such as manufacturing, agroindustry, and health, among others. One important aspect to clarify is that the sample size used in this project may appear to be small (i.e., this is a potential limitation). However, the sample size is relative to the model complexity, as well as to the a priori existence of the strong theory related to the instrument to be validated [49,50]. In other words, in our case, there is no solid a priori theory or a similar instrument (which is a limitation); for this reason, the sample was small. In addition, the said sample can be used for both the EFA and CFA, without losing reliability as part of the validation of our instrument. As part of future work, this proposed instrument should be further analyzed with other samples using only the CFA. Regarding the theoretical implications, our work presents evidence of an instrument with multiple ergonomic dimensions, which was not found in the literature concerning job satisfaction and occupational health. Moreover, regarding the practical implications, the results of this work can be used to promote and maintain higher degrees of physical, mental, and social well-being for workers as part of occupational health and job satisfaction.

Author Contributions: Conceptualization, V.M.R.-G. and I.O.-V.; methodology, V.M.R.-G. and J.A.L.-L.; software, V.M.R.-G., J.A.L.-L. and R.I.R.-G.; validation, V.M.R.-G., J.A.L.-L., R.V.-M. and R.I.R.-G.; formal analysis, V.M.R.-G., J.A.L.-L., S.V.-M. and P.G.-O.; investigation, V.M.R.-G., J.J.G.-O., J.A.L.-L., R.I.R.-G. and P.G.-O.; resources, V.M.R.-G., J.J.G.-O., J.A.L.-L., R.I.R.-G. and P.G.-O.; data curation, V.M.R.-G., J.A.L.-L., S.V.-M. and R.I.R.-G.; writing—original draft preparation, V.M.R.-G., J.A.L.-L., R.I.R.-G. and P.G.-O.; writing—review and editing, V.M.R.-G., J.A.L.-L., R.I.R.-G., R.V.-M. and P.G.-O.; visualization, V.M.R.-G. and J.A.L.-L.; supervision, V.M.R.-G. and I.O.-V.; project administration, V.M.R.-G.; funding acquisition, V.M.R.-G. All authors have read and agreed to the published version of the manuscript.

Funding: This research was funded by Universidad de Sonora.

Institutional Review Board Statement: This study was conducted in accordance with ethical regulations and approved by the Ethics Committee of Universidad de Sonora (202201NAV).

Informed Consent Statement: Informed consent was obtained from all subjects involved in the study.

Data Availability Statement: Not applicable.

Conflicts of Interest: The authors declare no conflict of interest.

References

1. Kumar, S.S. Job satisfaction-An overview. *Int. J. Res. Econ. Soc. Sci.* **2016**, *6*, 200–206.
2. Van Saane, N.; Sluiter, J.K.; Verbeek, J.H.A.M.; Frings-Dresen, M.H.W. Reliability and validity of instruments measuring job satisfaction—A systematic review. *Occup. Med.* **2003**, *53*, 191–200. [CrossRef] [PubMed]
3. Wright, T.A. The emergence of job satisfaction in organizational behavior: A historical overview of the dawn of job attitude research. *J. Manag. Hist.* **2006**, *12*, 262–277. [CrossRef]
4. Lange, T. Job satisfaction and implications for organizational sustainability: A resource efficiency perspective. *Sustainability* **2021**, *13*, 3794. [CrossRef]
5. Davidescu, A.A.M.; Apostu, S.A.; Paul, A.; Casuneanu, I. Work flexibility, job satisfaction, and job performance among romanian employees-Implications for sustainable human resource management. *Sustainability* **2020**, *12*, 6086. [CrossRef]
6. Jigjiddorj, S.; Zanabazar, A.; Jambal, T.; Semjid, B. Relationship Between Organizational Culture, Employee Satisfaction and Organizational Commitment. *SHS Web Conf.* **2021**, *90*, 02004. [CrossRef]

7. Buchbinder, S.B.; Wilson, M.; Melick, C.F.; Powe, N.R. Primary care physician job satisfaction and turnover. *Am. J. Manag. Care* **2001**, *7*, 701–713.

8. Lepold, A.; Tanzer, N.; Bregenzer, A.; Jiménez, P. The efficient measurement of job satisfaction: Facet-items versus facet scales. *Int. J. Environ. Res. Public Health* **2018**, *15*, 1362. [CrossRef]

9. Shepti, N.; Awaluddin, R. The Effect of Quality of Work Life, Organizational Culture and Job Satisfaction on Employee Engagement. Bina Bangsa. *Int. J. Buss. Manag.* **2021**, *1*, 158–165.

10. Ladebo, O.J. Effects of Work-related Attitudes on the Intention to Leave the Profession. *Educ. Manag. Adm. Leadersh.* **2005**, *33*, 355–369. [CrossRef]

11. Govea Andrade, K.; Zuñiga Briones, D. Clima Organizacional como Factor en la Satisfacción Laboral de una Empresa de Servicios. *Investig. Neg.* **2020**, *13*, 15–22. [CrossRef]

12. Mehrabi, J. Application of Six-Sigma in Educational Quality Management. *Procedia Soc. Behav. Sci.* **2012**, *47*, 1358–1362. [CrossRef]

13. Bauer, T.K. High Performance Workplace Practices and Job Satisfaction: Evidence from Europe. *SSRN Electron. J.* **2021**, *55*, 57–85. [CrossRef]

14. Nikolić, M.; Vukonjanski, J.; Nedeljković, M.; Hadžić, O.; Terek, E. The impact of internal communication satisfaction dimensions on job satisfaction dimensions and the moderating role of LMX. *Public Relat. Rev.* **2013**, *39*, 563–565. [CrossRef]

15. Bakar, N.A.; Radzali, N.A. Factors and job satisfaction dimension among academic staffs of public universities. *Int. J. Recent Technol. Eng.* **2019**, *8*, 427–430. [CrossRef]

16. Chavalitsakulchai, P.; Ohkubo, T.; Shahnavaz, H. A Model of Ergonomics Intervention in Industry: Case Study in Japan. *J. Hum. Ergol.* **1994**, *23*, 7–26. [CrossRef]

17. Kleiner, B.M. Macroergonomic Analysis and Design for Improved Safety and Quality Performance. *Int. J. Occup. Saf. Ergon.* **1999**, *5*, 217–245. [CrossRef]

18. Gomathi, K.; Rajini, G. Organizational ergonomics: Human engineering leading to employee well-being. *Int. J. Eng. Innov. Technol.* **2019**, *8*, 3744–3749. [CrossRef]

19. Carayon, P.; Hoonakker, P.; Haims, M.C. Participatory Ergonomics and Macroergonomic Organizational Questionnaire Surveys. *Proc. Hum. Factors Ergon. Soc. Annu. Meet* **2002**, *46*, 1351–1354. [CrossRef]

20. Haro, E.; Kleiner, B.M. Macroergonomics as an organizing process for systems safety. *Appl. Ergon.* **2008**, *39*, 450–458. [CrossRef]

21. Dalle Mura, M.; Dini, G. Optimizing ergonomics in assembly lines: A multi objective genetic algorithm. *CIRP J. Manuf. Sci. Technol.* **2019**, *27*, 31–45. [CrossRef]

22. M. Noor, M.H.H.; Raja Ghazilla, R.A. Physical ergonomics awareness in an offshore processing platform among Malaysian oil and gas workers. *Int. J. Occup. Saf. Ergon.* **2020**, *26*, 521–537. [CrossRef] [PubMed]

23. Kohli, A.; Sharma, A. The Critical Dimensions of Job Satisfaction of Academicians: An Empirical Analysis. *IUP J. Organ. Behav.* **2018**, *17*, 21–35.

24. McKeown, C. A Guide to Human Factors and Ergonomics. *Ergonomics* **2008**, *51*, 949–951. [CrossRef]

25. Sanjog, J.; Patel, T.; Karmakar, S. Occupational ergonomics research and applied contextual design implementation for an industrial shop-floor workstation. *Int. J. Ind. Ergon.* **2019**, *72*, 188–198. [CrossRef]

26. Tarka, P. An overview of structural equation modeling: Its beginnings, historical development, usefulness and controversies in the social sciences. *Qual. Quant.* **2018**, *52*, 313–354. [CrossRef]

27. Morrison, T.G.; Morrison, M.A.; McCutcheon, J.M. Best Practice Recommendations for Using Structural Equation Modelling in Psychological Research. *Psychology* **2017**, *08*, 1326–1341. [CrossRef]

28. Banerjee, T.; Banerjee, A.; Paul, E. *A Conceptual Overview of Structural Equation Modeling. IIMA Working Paper*; Indian Institute of Management: Indore, India, 2011; pp. 1–24.

29. Ballen, C.J.; Salehi, S. Mediation Analysis in Discipline-Based Education Research Using Structural Equation Modeling: Beyond "What Works" to Understand How It Works, and for Whom. *J. Microbiol. Biol. Educ.* **2021**, *22*, e00108-21. [CrossRef]

30. Leguina, A. A primer on partial least squares structural equation modeling (PLS-SEM). *Int. J. Res. Method Educ.* **2015**, *38*, 220–221. [CrossRef]

31. Soelaiman, N.F.; Ahmad, S.S.S.; Mohd, O.; Al Hakim, R.R.; Hidayah, H.A. Modeling the civil servant discipline in Indonesia: Partial least square-structural equation modeling approach. *Asean Int. J. Bus.* **2022**, *1*, 43–58. [CrossRef]

32. Afandi, M.; Zuleta, M.S.; Neolaka, A. Causative Correlation of Teacher's Motivation and Discipline in Banyumanik, Semarang City. *Int. J. Instr.* **2020**, *14*, 507–520. [CrossRef]

33. Watkins, M.W. Exploratory Factor Analysis: A Guide to Best Practice. *J. Black Psychol.* **2018**, *44*, 219–246. [CrossRef]

34. Yong, A.G.; Pearce, S. A Beginner's Guide to Factor Analysis: Focusing on Exploratory Factor Analysis. *Tutor. Quant. Methods Psychol.* **2013**, *9*, 79–94. [CrossRef]

35. Howard, M.C. A Review of Exploratory Factor Analysis Decisions and Overview of Current Practices: What We Are Doing and How Can We Improve? *Int. J. Hum. Comput. Interact.* **2016**, *32*, 51–62. [CrossRef]

36. Schmitt, T.A. Current methodological considerations in exploratory and confirmatory factor analysis. *J. Psychoeduc. Assess* **2011**, *29*, 304–321. [CrossRef]

37. Marsh, H.W.; Guo, J.; Dicke, T.; Parker, P.D.; Craven, R.G. Confirmatory Factor Analysis (CFA), Exploratory Structural Equation Modeling (ESEM), and Set-ESEM: Optimal Balance Between Goodness of Fit and Parsimony. *Multivar. Behav. Res.* **2020**, *55*, 102–119. [CrossRef]

38. Shea, B.J.; Grimshaw, J.M.; Wells, G.A.; Boers, M.; Andersson, N.; Hamel, C.; Ashley, C.P.; Peter, T.; David, M.; Bouter, L.M. Development of AMSTAR: A measurement tool to assess the methodological quality of systematic reviews. *BMC Med. Res. Methodol.* **2007**, *7*, 10. [CrossRef]
39. Yu, J.; Wu, Y. The Impact of Enforced Working from Home on Employee Job Satisfaction during COVID-19: An Event System Perspective. *Int. J. Environ. Res. Public Health* **2021**, *18*, 13207. [CrossRef]
40. Yu, X.; Zhao, Y.; Li, Y.; Hu, C.; Xu, H.; Zhao, X.; Huang, J. Factors Associated with Job Satisfaction of Frontline Medical Staff Fighting Against COVID-19: A Cross-Sectional Study in China. *Front. Public Health* **2020**, *8*, 426. [CrossRef]
41. Dymecka, J.; Filipkowski, J.; Machnik-Czerwik, A. Fear of COVID-19: Stress and job satisfaction among Polish doctors during the pandemic. *Adv. Psychiatry Neurol.* **2021**, *30*, 243–250. [CrossRef]
42. Taylor, J.M. Overview and illustration of Bayesian confirmatory factor analysis with ordinal indicators. Practical Assessment. *Res. Eval.* **2019**, *24*, 4. [CrossRef]
43. Markus, K.A. Principles and Practice of Structural Equation Modeling by Rex B. Kline. *Struct. Equ. Model.* **2012**, *19*, 509–512. [CrossRef]
44. Jöreskog, K.G. A general approach to confirmatory maximum likelihood factor analysis. *Psychometrika* **1969**, *34*, 183–202. [CrossRef]
45. Faller, H.; Kohlmann, T.; Zwingmann, C.; Maurischat, C. Exploratory and confirmatory factor analysis. *Rehabilitation* **2006**, *45*, 243–248. [CrossRef]
46. Peterson, R.A. Meta-analysis of Alpha Cronbach's Coefficient. *J. Consum. Res.* **2013**, *21*, 381–391. [CrossRef]
47. Miles, A.K.; Perrewé, P.L. The Relationship Between Person-Environment Fit, Control, and Strain: The Role of Ergonomic Work Design and Training. *J. Appl. Soc. Psychol.* **2011**, *41*, 729–772. [CrossRef]
48. Kabungaidze, T.; Mahlatshana, N.; Ngirande, H. The Impact of Job Satisfaction and Some Demographic Variables on Employee Turnover Intentions. *Int. J. Bus. Adm.* **2013**, *4*, 53–65. [CrossRef]
49. Izquierdo, I.; Olea, J.; Abad, F.J. Exploratory factor analysis in validation studies: Uses and recommendations. *Psicothema* **2014**, *26*, 395–400. [CrossRef]
50. Hurley, A.E.; Scandura, T.A.; Schriesheim, C.A.; Brannick, M.T.; Seers, A.; Vandenberg, R.J.; Williams, L.J. Exploratory and Confirmatory Factor Analysis: Guidelines, Issues, and Alternatives. *J. Organ. Behav.* **1997**, *18*, 667–683. [CrossRef]

International Journal of
*Environmental Research
and Public Health*

Article

Occupational Stress in Chinese Higher Education Institutions: A Case Study of Doctoral Supervisors

Xueyu Wang

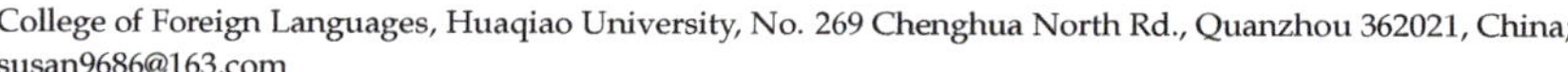

College of Foreign Languages, Huaqiao University, No. 269 Chenghua North Rd., Quanzhou 362021, China; susan9686@163.com

Abstract: This qualitative study is intended to explore the factors that contribute to the occupational stress suffered by Chinese doctoral supervisors and the kind of measures needed to effectively address the issue. Through purposive and snowballing sampling, 30 Chinese doctoral supervisors in different disciplines of natural science and social science were selected. A semi-structured interview protocol was used, and the data were analyzed based on grounded theory methodology. Chinese doctoral supervisors experienced varied stressors of nuanced nature, which could be categorized into two core categories, i.e., performance-appraisal-related factors and Ph.D. student-related factors, which were further divided into 18 subcategories and 10 higher-level categories. Chinese doctoral supervisors are under various sources of stress, corroborating with and reinforcing previous research findings in respect to occupational stress worldwide. Through the analysis of the stress triggers, suggestions are presented in regard to what mental health professionals and educational policy makers can do to address the issue of concern for doctoral supervisors.

Keywords: higher education; supervisors; distress; factors; grounded theory

Citation: Wang, X. Occupational Stress in Chinese Higher Education Institutions: A Case Study of Doctoral Supervisors. *Int. J. Environ. Res. Public Health* **2022**, *19*, 9503. https://doi.org/10.3390/ijerph19159503

Academic Editor: Masahito Fushimi

Received: 12 July 2022
Accepted: 29 July 2022
Published: 2 August 2022

Publisher's Note: MDPI stays neutral with regard to jurisdictional claims in published maps and institutional affiliations.

1. Introduction

Research worldwide reveals that occupational stress and workplace health has been a widespread epidemic for all genders and ages, and has become an issue of great concern over the last decade, both internationally and nationally [1]. This study adopts the definition of "stress" by the Canadian Mental Health Association, i.e., it is a reaction to a situation—it is not about the actual situation. In other words, people usually feel stressed when they think that the demands of the situation are greater than their resources to deal with that situation (https://www.ccohs.ca/oshanswers/psychosocial/stress.html, accessed on 18 December 2018). Given the value of work in this society, the amount of time spent at work and the current changes that are affecting the nature of work, it is not surprising that work stress appears to be increasing [2].

This is especially the case in educational institutions. Apart from various reports about student stress [3,4], a brief review of the literature shows the prevalence of stress, anxiety and depression among teaching staff. For instance, the 1998 and 2004 surveys in British universities revealed that, compared with other professions in the United Kingdom, the academic staff were experiencing higher levels of psychological distress, due to the increasing responsibilities and the high demands of their jobs, and there had been little betterment in the work standards of health and safety within six years [5]. Meanwhile, a study of the 1086 employees of a Quebec university in Canada found that the level of psychological distress reported from individuals was twice as high (40%) as that in a Quebec-wide sample (20%) [6], echoing the findings of the prevalence of occupational stress, anxiety and depression among Egyptian teachers in another study, due to such independent predictors for depression as inadequate salary, higher qualifications, workload, etc. [7].

In China, research also reveals that stress, which causes a chain of psychological problems, is not only epidemic among primary and secondary school teachers, closely attributing to gender, school types, marital status, physical health, economic situation, etc. [8], or professional pressure [9], but also widespread among university teachers, due to sense of value, sense of meaning, sense of identity and sense of helplessness [10] or due to scientific research performance pressure and their occupational stress and anxiety [11]. These findings, along with other reports, reveal a striking fact that in China, there is a tendency for academic scholars to die young, i.e., quite a few prominent scholars died unexpectedly in their prime, which to an extent indicates stress is prevalent among academia (https://www.zhihu.com/question/334804133, accessed on 26 July 2020).

Although a growing line of research worldwide has examined mental health problems in academic contexts and this undertaking has clearly yielded a range of important insights, and internationally, there is an increased emphasis on formal training, monitoring and accountability of doctoral supervisors (e.g., [12,13]), little in-depth research has ever been carried out to explore the kind of anxiety or stress that doctoral supervisors experience and why they suffer from stress of this kind or another, except for a limited number of studies which address, respectively, doctoral supervisors' emotional, professional and intellectual issues through personal, learning and institutional dimensions [14], and supervisors' anxiety or stress caused by the change in supervisory relationships (i.e., when a supervisor takes on a Ph.D. student previously supervised by another or has to hand over a student to another supervisor's care) [15].

Regretfully, no research has ever been carried out to explore the stress issue among Chinese doctoral supervisors. However, quite a few news reports about higher education in recent years have highlighted the importance of the stress issue among doctoral supervisors. For example, on 7 July 2019, a professor and doctoral supervisor at Tsinghua University, died at the age of 56 (http://www.bjnews.com.cn/edu/2019/07/09/600979.html, accessed on 9 July 2019); in October 2020, a female professor of history at Fudan University died at the age of 42 (https://new.qq.com/rain/a/20201012A0DZ1300, accessed on 12 October 2020). They are but a few of those scholars who died in their forties or fifties, so rumor has it that fifty is a doomed age for academics. Further, in 2016, a professor of literature and doctoral supervisor at Xiamen University, after being disqualified from being a supervisor due to his failure to comply with the university policy governing the recruitment process for Ph.D. students (http://news.sohu.com/20160224/n438383997.shtml, accessed on 24 February 2016), published an open letter proclaiming that he had decided to leave academia.

These cases could be just a tip of the iceberg. More cases and reports popping up time and again in social media highlight the psychological stress experienced by Chinese doctoral supervisors, attracting more concerns and attentions from the public. Doctoral supervisors are usually those limited individuals who enjoy a high-level reputation for their academic achievements in certain areas. For the public, the title of "supervisor" is not only a symbol of academic awards, but also a symbol of prominent social status. But few people could imagine that under this glory, these individuals' mental health is eroded bit by bit due to the overwhelming stress. The striking silence about the prevalence and determinants of mental health problems among Chinese doctoral supervisors necessitates a further study to address this gap in the doctoral education literature.

2. Materials and Methods

Research methods applied in previous studies range from quantitative to mixed approaches, and instruments are adopted for statistical analyses and calculation of stress levels (e.g., [6,7]). However, the purpose of this study is to explore the factors that contribute to the occupational stress suffered by Chinese doctoral supervisors. So qualitative research methods are appropriate for in-depth investigation on this issue, and among them, grounded theory is helpful to understand how and why Chinese doctoral supervisors are experiencing stress.

2.1. Research Design

Grounded theory (GT), a well-established methodological and qualitative approach for context-specific inductive theory building or an inductive enquiry that explains social processes in complex real-world contexts [16], is favored in this study, as it provides a major contribution to the generation of emergent theory when there is little known about a particular phenomenon. GT is applied based on the Strauss and Corbin approach [17]. As such, it is necessary to conduct grounded theory, in order to understand the stress-related experiences of doctoral supervisors and establish some countermeasures.

A pilot study was conducted to serve such purposes as adding value and credibility to the semi-structured interviews to be used in the major study, avoiding inefficient theoretical sampling or even erroneous purposeful sampling and providing remedial loops, or effecting changes of focus, guidance to improve data collection instruments, and/or informing theoretical sampling [18], or avoiding certain ethical problems [19], among others. Specifically, initial interview questions were first designed pertaining to the stressors that supervisors may endorse (Table 1), then open-ended questions were emailed for expert reviews in the interview protocol. The broad and general initial interview questions were revised, which was based on the suggestions from the experts, in order to prepare for defining and clarifying the research theme in the pilot testing. After that, participants were recruited from the selected universities and the pilot interview was conducted. Purposeful sampling was adopted in the selection of the informants, so that the participants in the pilot interview shared as similar criteria as possible to the group of participants in the major study [20,21]. Two doctoral supervisors, in their 40s and 50s, were interviewed in their offices in October 2021. The two interviews, including the social conversation to build rapport as suggested by Jacob and Ferguson [22], ranged in time between approximately 35 min to 45 min and were recorded using a smart phone. Both of them were informed of the research goal and were willing to participate in it.

Table 1. Interview guide.

Order	Questions
1	What comes to your mind when "stress" is mentioned?
2	Do you think stress is prevalent among doctoral supervisors? If so, give examples;
3	During your life time as a doctoral supervisor, have you ever suffered from any stress? If so, what are the main stressors?
4	Have you received any training with respect to addressing mental health issues before you are elected as a doctoral supervisor? Is the training necessary? What, if any, are the positive aspects of the training?
5	In respect to the stressors, what suggestions would you propose to alleviate the distressing symptoms among supervisors?
6	Is there anything else you would like to add?

Upon completion of the initial interview, the data were transcribed and analyzed to deduce themes and concepts. The analysis of research participants' response to the questions laid a basis for the questions to be used in the follow-up interviews. As themes emerged, the interview questions for the second round of interviews were narrowed down and focused on the existing themes. In this way, new themes emerged while the existing themes were confirmed. So, the follow-up interviews were conducted to confirm or clarify emergent themes, and meanwhile, to explore any new information. The pilot study helped to test the appropriateness of the interview questions, and adjust the research plan, identify the possible problems, obtain experience and build rapport in the follow-up in-depth, semi-structured interviews in the major study.

2.2. Participants and Ethical Considerations

Throughout the process of participant recruitment, purposeful and snowball sampling was used, i.e., early recruits led to the later recruits. The participants consisted of 15 men and 15 women who ranged in age from 33 to 62 at the time of the interview, came from northwestern, southwestern, central, northeastern and southeastern areas of China, and majored in varied disciplines of natural science and social science. Except one who had a Master's degree, the rest of the participants had doctoral degrees. Participant selection, data collection and data analysis continued until theoretical saturation reached. Data collection ceased after 25 interviews, as it was clear that no new themes emerged; but 5 additional participants were interviewed to reduce the chance of missed themes and ensure that data saturation was achieved.

Prior to the commencement of the interview, an initial explanatory statement was sent via WeChat to identified supervisors who were informed about the purpose of the study, the length of the taped interview (approximately 35 to 45 min in duration), the voluntary and confidential nature of the interview (i.e., any personally identifiable information, such as their names and affiliations, would be removed or changed before research findings were shared with other researchers or results were made public), the way of answering the study instruments, their right of withdrawal or termination (i.e., they were free to withdraw from the research at any point after initially consenting to participate) and written consent for participation was sought from all participants. Further, official approval was sought from pertinent authority, so as to comply with pertinent codes of ethics (e.g., [23–25]).

2.3. Interview and Data Collection

At the outset of the interviews, participants were reassured of anonymity and confidentiality of the interview, helping to build trust [26], and semi-structured in-depth interviews were used for its varied functions, such as encouraging depth and vitality [27,28] or soliciting new concepts to emerge [29]. Due to the novel coronavirus disease (COVID-19), interviews were conducted via WeChat during the past few months. After a brief introduction to his/her personal information, each participant answered the open-ended questions, which facilitated a free flow of ideas from the respondent and generated information-rich data. Meanwhile, a clinical method for interviewing was adopted [30]. Thus, the participants were requested to elaborate on a statement if it was vague or ambiguous, so as to ensure the clarity of certain concepts.

Each interview was conducted in Chinese and audio recorded by the in-built audio recording software of WeChat. Each participant was assigned a code number to avoid revealing his or her identity. A professional transcriptionist was hired to transcribe the audio files into Mandarin Chinese, and each interview transcript was under a first-pass verification, i.e., verified by each participant for accuracy and validity, then the second-pass review was conducted by the interviewer using the original audio files and anonymized, formatted transcripts, to complete a final check of content accuracy and then adjust the level of transcription for analysis. After the transcripts were double checked, they were translated into English.

In qualitative research, interview translations always aim for semantic equivalence at a minimum and aspire to conceptual equivalence [31]. To validate the translation quality and demonstrate the functionally equivalent translation, the back-translation method was adopted, as suggested by Brislin [32].

After verifying the trustworthiness or credibility and increasing the accuracy of the study through triangulation, i.e., participant checking, self-reflection and peer review [33], the transcripts and the translations thereof were stored and analyzed side by side. A format came into being that helped the research team to discuss the transcripts and interact with the original and translated versions [34]. Data coding was conducted by language-congruent researchers in the source language, adjacent to the same data in the English translation, as recommended by Olson [35].

2.4. Data Analysis Method

Based on the questions in the interview guide, verbatim transcripts were read and analyzed repeatedly to identify themes related to stress, then filtered and selected through the constant comparative method. The whole process involved three phases of coding (i.e., open, axial and selective coding) [17]. First, the stress factors experienced by supervisors were identified by dismantling and reviewing the data collected through open coding to form codes, then the codes were categorized. Second, all the higher order codes were merged, and data were grouped and linked to each other through axial coding. The relationships between categories were organized into a paradigm model that includes, causal conditions, contextual conditions, interventional conditions, action-interaction and consequences. Third, the core categories were identified through selective coding, thus leading to a substantial theory that could explain the stressors experienced by Chinese doctoral supervisors (Table 2).

Table 2. Coding of identified occupational stress suffered by Chinese doctoral supervisors.

Paradigm	Categories	Subcategories
Causal conditions	Demanding academic requirements	Publication of 3 or more articles in key journals within 3 or 4 years
		Hosting a research program at ministerial level or above
	Heavy teaching workload	Annual workloads ranging from 300 to 340 credit hours in two semesters
	Various student-related issues	Student graduation with article publication and thesis completion
		Negative relationship between supervisors and Ph.D. students
		Constraint of new Ph.D. student recruitment
Central phenomenon	Experience of stress	Negative emotions (dissatisfaction, frustration, exhaustion, loss of face etc.)
Contextual conditions	Policies and regulations in higher education	Unreasonable performance appraisal
		Complex student-related issues
Intervening conditions	Personal performance appraisal	Difficulties in article publication
		Requirements for awards and programs
	Doctoral student-related issues	Tensed relationship between supervisors and Ph.D. students
		Various *guanxi* involved in the recruitment of Ph.D. students (*Guanxi* (关系), in Chinese culture, means the system of social networks and influential relationships which facilitate business and other dealings. *Guanxi* is often translated as "connections", "relationships" or "networks".).
Action/Interaction strategies	Working with long hours	Academic research (article publication, hosting/winning research programs and awards)
		Teaching workload
	Self-compromise	Dealing with student-related issues
Consequences	Experiencing various stressors	A single stressor
		Combination of a set of stressors

3. Results

The participants revealed a variety of factors contributing to this widespread stress among Chinese doctoral supervisors. Nevertheless, the GT emerging from the data represented the nuanced nature of participants' experiences with stress. Categorization of the interview data through open coding yielded 18 subcategories and 10 higher-level categories. The relations of categories were linked through axis coding according to the paradigm model (Figure 1). Data analysis was found to comprise two core categories, i.e., performance appraisal and Ph.D. student-related issues.

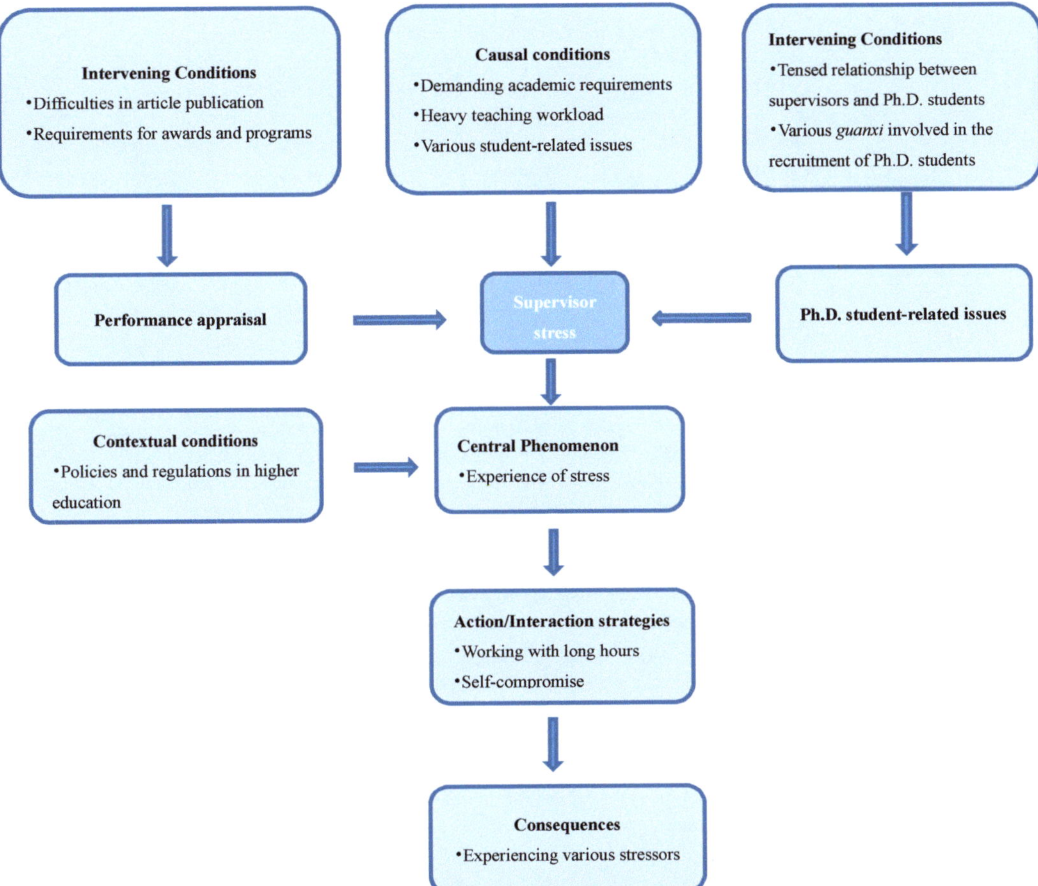

Figure 1. Paradigm model of occupational stress suffered by Chinese doctoral supervisors.

3.1. Causal Conditions

The participants were found to be engaged in "demanding academic requirements". First, it was extremely tough, if not totally impossible, to publish three or more articles in such designated journals indexed by SCI, SSCI or CSSCI (the abbreviation of Chinese Social Sciences Citation Index "中文社会科学引文索引", herein after referred as "*hexin* journals", meaning "key journals") during the contract term of three or four years. Second, the requirement to host funded research program at ministerial level or above increased the stress for the participants, although for some participants, it was an alternative to article publication requirement.

" . . . You know, almost all universities in China developed a Guide for Journal Category. Category D1 journals include the rest of journals covered in CSSCI except for Category A, B and C journals. My university adopted the similar journal categorization system that CSSCI journals are categorized into A1, A2, A3 . . . ; B1, B2, B3 . . . To get three articles published in these Category C journals, or two articles in Category C journals and two articles in Category D1 journals in three years, is absolutely a fable, at least to me." (Participant 12)

" . . . Winning any award or hosting a *NSSFC* (National Social Science Fund of China) or *NSFC* (National Natural Science Foundation of China) is tantamount to finding a needle in a haystack." (Participant 3)

The interview data also revealed that the supervisors had to undertake "heavy teaching workload". Different from the credit hour adopted in western countries, the workload (工作量), as devised by the Human Resources of varied Chinese universities, comprises the number of equivalent 40 (45, or 50)-min lecture periods and credits for supervising laboratories or supervising graduates per semester. Some universities require that supervisors independently teach at least one undergraduate level course and supervise a certain number of postgraduates per year. The annual workloads of the participants ranged from 300 to 340 credit hours in two semesters, which is equivalent to teaching five to six courses per semester, namely, 10 or 12 credit hours per week for 16–18 weeks. A minimum of their annual workloads, e.g., 120 credit hours of undergraduate-level teaching is required, depending on their respective salary range and position grade they have contracted with their universities.

"As a full-time professional, the workload is too much for me. It is tough to strike a balance between academic research and teaching." (Participant 4)

Further, the supervisors had the end-of-course teaching evaluations by students via a Management Information System at the end of each semester. This system is run by the University, also called teacher performance appraisal system (TPA). It is a 100-point scale questionnaire. Students are instructed to log in to fill out the TPA form online. The evaluation ratings for teachers should reach a minimum score of 90 for teachers to satisfy the requirements set for the position of their respective salary grade.

" . . . student appraisal is always affected by some unexpected or irritating factors . . . e.g., low scores will result in retaliation from students . . . " (Participant 2)

As for "various student-related issues", the participants reported that student graduation with article publication and thesis completion was the key source of their stress. Moreover, the supervisor–student relationship was an unexceptional trigger for stress. With regards to new Ph.D. student recruitment, not only the unspoken rule or *guanxi*, but the number of recruitments required would be stressful.

"If a student can't publish an article in a *hexin* journal as required by the University, he/she can't graduate on time. Further, he or she must compose a thesis (either a monograph or compilation), undergo external review of thesis, and pass the public defense of dissertation for Ph.D." (Participant 13)

" . . . Some of my students seldom clean up laboratory glassware or other instruments after they finish experiments, and reluctantly comply with experiment regulations or protocols, although I remind them time and again about the importance of following the rules." (Participant 11)

" . . . Sometimes, it is really hard to strike a balance between those varied *guanxi*, especially when I have some adoring students who decide to pursue a doctoral degree. As such, I often maneuver into a peculiar dilemma as to which student to recruit." (Participant 16)

3.2. Central Phenomenon

All the participants shared "negative emotions" throughout the interview process. The pervading gloominess, sullenness or stress was obvious or pervasive when they mentioned that their articles had been repeatedly rejected by varied journals. Some participants were wondering whether they were psychologically healthy, saying that they could not imagine gloating over other people's failure. In a sense, article publication involves losing or gaining face.

" … one of my daily routines is to ask whether my colleagues have got article rejected, accepted or published. Negative response would console or excite me, while positive response would make me jealous and more distressful." (Participant 15)

" … Supervisors are supposed to have publications in high impact factor journals, and to inspire the students as a role model and command respect." (Participant 6)

All participants also remarked that they were always in a paradox, i.e., if their grant proposals were rejected, they would feel heartbroken, but being granted a funding program was nothing more than a short-lived pleasure of success, or just the beginning of another disaster.

" … I would always be on tenterhooks before program approval is published. When your application is rejected, you would think the sky is about to fall down." (Participant 5)

In all, 8 out of 30 participants remarked that their workload was rather heavy and exhausting. As for teaching evaluations, although 21 participants seemed not bothered by it at all, 9 participants deemed it as a major stressor.

"The assigned workload makes me worn out, and I easily get irritated even over trifles in daily life." (Participant 4)

" … you know, a low rating for your teaching not only means loss of money, but more importantly, loss of face." (Participant 18)

During the interviews, all the participants shared the view that thesis and article publication issues were key sources of stress, for failure to meet these requirements not only jeopardizes students' future (namely, rejected doctorates), but also exerts significant negative impacts on their supervisors, e.g., illegitimacy to recruit new Ph.D. students and loss of face, among others.

A total of 13 out of 30 participants mentioned their stressful experience with students. Half of the participants proclaimed to have been stressed by the unspoken rule or *guanxi* issue in the process of recruiting new Ph.D. students.

3.3. Contextual Conditions

The participants reported that their stress was mostly attributed to "policies and regulations in higher education", one of which was performance appraisal. As a rating mechanism, performance appraisal is employed by almost all universities and institutions of higher education. In this way, job performance of an employee is evaluated in terms of quality, quantity, cost and time, and human resources development provides continued push for personnel accountability and competency and effective schooling, and determines who needs what training, and who will be promoted, demoted, retained or fired [36]. As dictated by the participants, performance appraisal, as a major stressor, consisted of such rating factors as article publication, research program and teaching.

"It is quite unreasonable to publish 3 or more articles in key journals, or host a research program at ministerial level or above during 4 years. Moreover, I have to undertake 340 workloads in two semesters. Otherwise I will be degraded or removed from the current post." (Participant 17)

According to the supervisor's guide promulgated by most universities, supervisors should have adequate funding to allow the research to progress (this is especially the case for supervisors of doctoral students in science and engineering), and having adequate funding (e.g., RMB 200,000 or RMB 400,000) was the prerequisite for new Ph.D. student recruitment. Consequently, supervisors were desperate for grant programs, availing every effort to draw funds from all sorts of corners to support their students' research.

"Hosting a NSSFC or NSFC or winning those awards not only helps satisfy the requirements for my salary range and position grade, but also provides additional funding to take on new students." (Participant 21)

As for student-related issues, student graduation comes first. In China, the total duration for completion of degree requirements and submission of thesis for external evaluation is three regular years, or a maximum of six years, as of the date of admission as a Ph.D. student, during which time each candidate shall complete required credits in a subject, and publish, out of his or her research work, at least one research publication (letter of acceptance included) in a university approved *hexin* journal of his or her research field. Further, he or she should accomplish a thesis, which should make a distinct contribution to knowledge and show ability to conduct original investigations and to test ideas whether of his own or of others, undergo external review of the thesis and pass an oral defense.

> "If my students can't meet the requirements, they have to extend their timeline of graduation. It will affect the recruitment of new doctoral students. What's worse, I will be disqualified from being a supervisor." (Participant 23)

In addition, universities are prone to adopt a policy in favor of students whenever there is supervisor–student confrontation or conflict, as evidenced by numerous cases where Ph.D. students suffered psychological stress, or committed suicide due to pressure from study, but their supervisors were criticized for pressing them too much. As remarked by most participants, supervisors were supposed to take responsibility for everything ranging from Ph.D. students' daily lives to their graduation, otherwise they would be criticized for omission. As a result, supervisors dare not blame or criticize students, although students are lazy and refuse to follow the instructions or abide by certain academic ethics.

> " ... to criticize students is to ask for trouble ... when a student broke down an expensive instrument due to non-compliance with laboratory regulations, I was irritated. But the most stressful thing was that I could not claim compensation, lest it should cause any stress to him and I get punished in turn. Today is yesterday's pupil." (Participant 11)

In terms of the recruitment of new doctoral students, *guanxi* plays a very important role in it.

> "I would like to recruit older candidates from the legal profession for their rich experience in law, but could not always be away from worldly wisdom." (Participant 19)

Further, Participants 10, 21 and 27 mentioned the number of Ph.D. students required to recruit was a stressor. The Ministry of Education sets a limit or quota for Ph.D. student recruitment in different subjects or disciplines for each year. As such, in some subjects, a supervisor may recruit one or two doctoral students per year, while in other subjects, two or three supervisors have to share one doctoral student each year.

3.4. Intervening Conditions

The participants were found to be faced with various stressors, including "personal performance appraisal" and "doctoral student-related issues". All the participants had difficulties in publishing articles in key journals within the term of performance appraisal, usually 3 or 4 years.

> "The publication threshold for me is 3 articles in Category C journals. Of the 500 journals covered in the CSSCI database, only 190 journals fall into Category C and above. But of the 190 journals, only 10 journals fall within my research field or discipline ... You may say that there are some SSCI indexed journals ... you know, it's even harder, for you have to compete with academics worldwide. What's worse, only the top 25% of SSCI indexed journals in one particular discipline are deemed as Category C, in accordance with the Journal Category Guideline devised by my university." (Participant 7)

Apart from publishing the designated number of articles, the participants reported that it was tough to meet the requirements to win awards or host research programs at ministerial level or above, from the beginning of application to the deadline.

> " ... it is one thing to be great at your job, but quite another to win any of these awards ... you must have good connections, otherwise your application

cannot even go beyond the university. You know, candidates for these awards are firstly recommended by the schools to the university, then recommended to the provincial or ministerial department for deliberation." (Participant 25)

" ... you have to compose a final performance report which adheres to the Grant Program Guideline, then submit it to the funding authority before the program is due, failing which, you cannot apply for any research program for a couple of years ... " (Participant 8)

The participants also shared the idea that a tensed supervisor–student relationship was a stressor, including:

(a)   Students' complaints about supervisor bias or omission,
(b)   Students' low caliber or non-commitment to research (including indolence, lack of necessary expertise in carrying on experiments, plagiarism, misattribution, dispute over co-authorship, noncompliance with academic ethics, etc.),
(c)   Students' request to change supervisors due to dislike for each other, and/or
(d)   other confrontations or noncooperation in relation to their study or research programs.

"Some students complain about my omission in their article publication, claiming that coloration, availability of funds and facilitation in article publication are the key features that must be contained in an active and productive Ph.D. supervisor ... it seems as if I have been grossly at fault for their manuscript rejection ... " (Participant 9)

One more stressor is related to various *guanxi* involved in the recruitment of new doctoral students.

" ... if an applicant finds that someone (e.g., a colleague, family member, or friend) is well connected to me, the applicant will normally request the contact(s) to refer or introduce him to me, or the contact(s) would ask me to do the applicant a special favor. Where there are two or more applicants, more contacts would come to me." (Participant 14)

3.5. Action/Interaction Strategies

The participants reported that in order to meet with the requirements of performance appraisal, they had to "work with long hours". They devoted nearly all their time and energy to academic research and teaching, wondering all day how to deliver excellent on-the-job performance and fulfill their on-the-job responsibilities, so much so that they spent less time on vacation or other leisure activities. Apart from article publication, they had to strive to succeed in:

(a)   Hosting a NSSFC or NSFC
(b)   Winning an *Award for Scholarly Achievement* (科研成果奖) (at least the Third prize) at national, ministerial or provincial level,
(c)   Winning an *Award for Teaching Achievement* (教学成果奖) (at least the Third prize) at Ministerial or Provincial level, or
(d)   Becoming a *Talent Excellence Program trainee* (优秀人才项目) at ministerial or provincial level.

" ... (being granted a funding program)what follows is long-lived torture of tedious research work, day and night, in order to produce either books, monographs, peer-reviewed articles, e-books, digital materials, experiment reports, translations with annotations or a critical apparatus, or critical editions resulting from previous research, before undergoing a review by relevant authority comprising of peer scholars." (Participant 7)

"I have to undertake 340 workloads in two semesters, and teaching students at undergraduate level shall account for 35% (i.e., 120 workloads) of the total workload. I try to strike a balance between academic research and teaching, always staying up late." (Participant 26)

While dealing with student-related issues, most participants claimed that they were liable in opting for "self-compromise", in order to avoid trouble or reduce the conflicts between supervisors and students.

> " ... Once, an experiment instrument worth ¥300,000 broke down for the student failed to comply with the obligations specified in the operation protocol. What stresses me out most is I can't claim compensation, lest it should cause any stress to him and I get punished in turn." (Participant 11)

> " ... this year, I have to share one doctoral student with a colleague. In the end, I stroke a deal with my colleague, i.e., I gave my half quota to him, so this student is under his supervision this year, and he promised to give me his half quota the following year ... this means I can recruit one student every two years." (Participant 10)

3.6. Consequences

It is noteworthy that despite their divergent perspectives as to age, gender, length of time and experiences in supervising students, etc., the participants "experienced various stressors", that is, they shared the same or similar psychological stress as an experience associated with their lives as doctoral supervisors; although some were stressed by a single factor, others were stressed by the combination of a set of factors.

> "Sometimes I feel frustrated, however, I make full use of my spare time to pursue projects that embody exceptional research, rigorous analysis, and clear writing, carefully and smartly articulate the projects' value to humanities scholars or general audiences, so that I might edge past the fierce competition and be granted the project." (Participant 22)

> " ... if I get a low rating for my teaching, I will fail the performance appraisal, resulting in deduction of annual salary, or demotion to a position of a lower salary grade as a punishment ... " (Participant 2)

> " ... having no students to supervise not only impairs the conduction of experiment, but also greatly impacts my performance appraisal." (Participant 19)

4. Discussion

The GT of supervisor stress, which was developed from the results of this study, indicates that supervisor stress is a multifaceted process inflicted by the influences of varied factors, individually or collectively, and their current stress at the time of the interview represents a result of the ongoing combination of value system, social norms and/or educational policies.

Apprehension at performance appraisal failure

Apprehension or angst at performance appraisal failure was viewed by all the participants as the most important trigger for their stress, although the subcategories mattered differently to the participants. Admittedly, it is rather difficult, if not altogether impossible, to publish three or more articles in *hexin* journals, receive funding (from national or provincial governments) for research projects and undertake 300–400 workloads simultaneously. After all, performance appraisal failure entailed not only disqualification for recruiting new Ph.D. students, but also reduced income and loss of dignity or face. This finding corroborates the findings of the most recent DFE (Department for Education) teacher workload survey that heavy workload is one of the biggest threats to teacher recruitment and retention and the level of workload is generating stress in schools [37].

The reason for imposing such harsh performance appraisal standards is closely tied to the tertiary education reforms.

One of the reforms is the scale expansion of Ph.D. students as a countermeasure to release the burden of unemployment rate among college graduates that has been rising in the past few years. The boosting enrollment means that supervisors have to supervise a growing number of Ph.D. students, which consequently contributes to the double dilemma arising from the apprehension at student graduation. Indeed, Chinese supervisors are carrying a supervision load of 5.77 doctorate candidates, much higher than the recommended threshold of 3 Ph.D. and 5 Master's students in any given academic year internationally. Theoretically, the more students they supervise, the more stressful they would find it. In contrast, in some countries, the average academic has a maximum potential of training 27 Ph.D. students in a 30-year career [38]. As more and more universities around the world graduate ever-increasing numbers of students with doctoral degrees, governments are beginning to ask if it is time to slow the production line [39].

The other reform is the Double First-Class Construction (*shuang yi liu*, 双一流, is short for "The World First-Class University and First-Class Academic Discipline" (Chinese: 世界一流大学和一流学科) Construction). This plan represents a new way of ranking universities in China. Up to date, according to the lists issued by the Chinese Ministry of Education, Double First-Class Universities consist of Class A (including 36 universities), Class B (including 6 universities), and Double Class discipline universities (including 95 universities). It was a tertiary education development initiative designed by the Chinese government in 2015 to develop elite Chinese universities and individual faculty departments into world-class institutions by the end of 2050. Currently, global university rankings attract considerable attention, and such rankings may cause universities to prioritize activities and outcomes that will have a positive effect in their ranking position [40]. The ranking systems affect universities globally [41]. To be shortlisted as a candidate of the Double First-Class Construction not only means being promoted to fame and a better student pool, but also to larger financial subsides and a higher income. As such, when designing a performance appraisal system, almost all universities set high academic and teaching standards, so that the shortlisted universities can maintain their places, while the universities that have not been shortlisted have a chance to enter. According to the Ministry of Education, the list is the result of competitive selection, expert evaluation, government assessment and periodic screening.

All the participants believed the performance appraisal was unreasonable, corroborating other findings that performance appraisal triggers nothing but dread and apprehension and is favored by no one except employers [42,43]. The process may conceal the decision-making and communication function of performance appraisal in organizations, despite the heightened view of appraisal as a measurement tool [44]. The number of publications or a single score on a rating form does not sufficiently measure the contributions of faculty.

Apprehension resulting from student-related issues

Of the student-related issues, apprehension at failure to graduation on time was a definite stressor for Ph.D. students. After all, to publish one or two articles in *hexin* journals is tough for supervisors, let alone Ph.D. students, corroborating other findings that doctoral students face high and potentially strenuous demands [45], and article publication is a major stressor for Ph.D. students [3].

Meanwhile, this is also a major stressor for supervisors, because supervisors' recruitment of new Ph.D. students and performance appraisal would be significantly impaired. The POP (publish or perish) practice is prevalent worldwide, obliging supervisors and Ph.D. students to spend more time scrambling to publish whatever they can get into print [46]. Since this practice entails inevitable ambivalence and resistance among doctoral supervisors and candidates in respect to the place of publication in doctoral work [47] and "mandating publications for graduation places a poor metric on Ph.D. students' skills and detrimental effects on Ph.D. training … " [48], it is suggested that the POP practice be abolished [49].

In addition, thesis writing is a stressor which not only increases stress for Ph.D. students [50,51], but also causes significant harm or severe emotional distress to supervisors. After all, the qualities of doctoral dissertations and graduation rates are directly related to the Ph.D. student recruitment and performance appraisal of supervisors.

The relationship between supervisors and Ph.D. students is a complex one. The tensed supervisor–student relationship, whether personal or professional, also makes for a major stressor. The interviews revealed that such factors as students' lack of interests and skill sets, non-cooperation, hostile conflict, mutual dissatisfaction and defiance of academic ethics, may result in tensions and stress to some supervisors, as evidenced by some news report about plagiarism allegation against two Ph.D. students whose supervisors have been temporarily suspended or permanently removed from the Supervisor Register [52]. These are just a few. Stories such as this pop up in the news regularly. For example, Jinan University, Anhui University and Fudan University accused some Ph.D. graduates of plagiarism, and supervisors of those graduates were suspended temporarily or removed from the Supervisor Register (https://zhuanlan.zhihu.com/p/46609954, accessed on 12 October 2018, http://news.sohu.com/20160322/n441458886.shtml, accessed on 22 March 2016 and https://lx.huanqiu.com/article/9CaKrnJZX3P, accessed on 22 January 2017, respectively).

The intricacy of the supervisor–student relationship corroborates with other findings that the supervisor–student relationship may be in part comparable to the one between the physician and his/her patient, and at some point of the journey, develop different expectations of one another [53], or exert negative effects on the supervisors' reputation, corroborating the findings that academic dishonesty, especially plagiarism, is a global problem that has bedeviled academia and has been viewed as unethical and immoral intellectual thievery that could negatively impact on not only the repute of an academic institution, but the prosperity of a society [54].

Factors that partly contribute to the tensed relationship can be summed up as follows:

(1) Students' failure in understanding what a doctoral degree means. A Ph.D. is the highest academic degree in China, which requires the command of more specialized academic knowledge in a certain field. Regretfully, some Ph.D. students have a lack of commitment to academic research, or not psychologically and professionally ready for the doctoral study. A survey conducted by *Nature* also reveals the turbulent nature of doctoral research [55]. Ironically however, a Ph.D. is increasingly becoming a springboard to a higher social status or prestige and greater power or resources. Actually, not every Ph.D. student fully understands that the word "philosophy" signifies an individual who has a passion of wisdom and a clear field of academic interest, and has achieved a comprehensive general education in the fundamental issues of the present world. Some of them merely focus on the final crossing of a symbolic boundary represented by the completed thesis and the viva, ignoring the fact that they should perform a "rite of passage" throughout their research education [56]. As such, the above-mentioned factors (such as indolence, noncompliance and noncooperation) are pervasive and consequently contribute to supervisor stress.

A recent blog published by a professor at Fudan University may represent the voice of all the stressed supervisors in China. The professor taunted about the decline in the competencies of postgraduates and Ph.D. students in China and called on his Ph.D. students to " … understand the stress and anxiety of 'an advanced animal' (i.e., the professor himself)"(https://baijiahao.baidu.com/s?id=1625877107917853133&wfr=spider&for=pc, accessed on 19 February 2019). Coincidentally, a professor at Shanghai Jiaotong University, who usually revised students' experiment protocols or articles until two o'clock in the morning, was greatly irritated by a student laziness or lack of academic expertise, and posted in a WeChat Group some sharp words such as "rubbish or garbage, shit" to address the students' experimental findings. Subsequently, the university delivered to him an official notice of suspension, required him to make a public apology to the student and issued a notice of criticism circulated within the department. These cases, among others, reveal in part the stress, anxiety, helplessness and apprehension confronting supervisors who are under significant pressure to complete some major yet urgent projects. So, if students have no clear conception about the doctorate, the stress will be there waiting for supervisors.

(2) Lack of training on the part of supervisors. Regretfully, in China, there is no mandatory requirement for formal training or monitoring and accountability of doctoral

supervisors, except for the obligatory academic requirements, for example, doctoral supervisors should have graduated with doctoral degrees, have published a required number of articles in *hexin* journals, or have been granted research programs at national or provincial level, as evidenced by the Ph.D. Supervision Guide promulgated by most universities in China. Actually, supervisors, especially the newly elected supervisors, undertake no such training prior to the commencement of doctoral supervision. As a result, supervisors are not completely prepared for the role of supervisor and to have insight into the responsibilities that it entails. After all, the supervisor–student relationship varies, while supervision training will make them aware of their roles and responsibilities.

Proposed suggestions

Anyway, the findings reveal the prevalence of stress among Chinese supervisors. Ironically, few people are aware that scholars are at greater risk of stress-related illness than police, medics and local authority staff [57]. At a time when student mental health has become the focal concern of most universities, it is surprising that staff well-being has been attracting the least attention [58].

Thus, it is high time that relevant policies were updated. Solutions proposed for the issue include:

(1) Clarifying respective rights and responsibilities

There is no doubt that supervisors should strictly abide by ethics and codes of practice related to their positions, and are highly expected to promote knowledge development and create a research culture during the research process [59]. However, emphasis should be placed on the norms and the innovations of Ph.D. students' academic research, such as the standards of citing references against plagiarism and the development of original ideas. Thus, it is indispensable for universities and institutions to draw up the specific guidelines for enrollment, supervision and assessment. Supervisors should be authorized to deal with Ph.D. students' academic performance in a more supportive and flexible way. Further, awards will go to supervisors to enhance their sense of achievement and develop the notion of "showing respect to teachers and promoting education (*zun shi zhong jiao* 尊师重教)".

(2) Establishing a full-fledged performance appraisal system

Nearly all the participants criticized the current policy makers for ignoring the fact that article publication was beyond their control, and suggested that university management adopt more flexible appraisal approaches. Fortunately, the Ministry of Education of China has been committed to the reform of educational assessment, one of the measures being to remove the "five-overemphasis". On November 8, 2018, the Ministry of Education carried out a campaign against "five-overemphasis" (破 "五唯"), that is, the overemphasis on articles, positions, professional titles, certificates and awards ("唯论文,唯帽子,唯职称,唯学历,唯奖项"). Its purpose is to further push the reform of systems within higher education, improve the mechanism of moral education("立德树人"), reverse the unreasonable evaluation guidance, operate on the appraisal system based on representative works and value the qualities and the influence of the works. In this way, multi-level performance appraisal is on trial at universities and colleges, including teaching, academic research and service, just as Participant 1 suggested that more indicators should be covered in the performance appraisal system, such as the social service of delivering public lectures or hosting hot-line meetings to address psychological issues related to the novel coronavirus disease. The reform values the evaluation of the teaching process and social service based on the annual assessment. The appraisal system not only covers self-assessment from supervisors, but the appraisal from their peers, Ph.D. students and the administrative staff. All these measures can serve to foster increases in personal knowledge of subject matter and effective methods for delivering the knowledge to students, to increase teacher confidence and, in turn, professional competence and to help teachers reach their potential [36]. After all, it is not the sole objective of the performance appraisal to determine who will be promoted, demoted, retained or fired, for this objective will only dilute and weaken the clarity and validity of any appraisal system [60].

(3) Adopting a more reasonable or flexible Ph.D. recruitment and education policy

This includes introducing a mechanism for changing supervisors when a confrontation between a Ph.D. student and his or her supervisor arises, and adopting an *Application and Verification System* (申请-审核制) under which Ph.D. candidates shall present their works (including articles published in key journals, programs hosted during their graduate study and research proposals) and show their personalities (e.g., independence, commitment to academic research, team work spirit), to convince their supervisor committee before they are admitted as Ph.D. students, and this system allows supervisors to dismiss disqualified Ph.D. students after due process, for example.

As to the enrollment policy reform, it is advisable to refer to the results of performance appraisal as an important decisive factor in the annual enrollment decision. Outstanding supervisors and teams are encouraged to share their experience and take the lead in Ph.D. education. In this way, improved student outcomes (i.e., qualified Ph.D. graduates) and fulfillment of the objectives of Ph.D. education could be ensured.

(4) Designing and implementing a supervisor training program

It is essential to establish a three-level (i.e., national, provincial and university level) training program, and organize pre-service and in-service training and assessment for supervisors each year. More measures are created to make the training fruitful, e.g., the application of moral education, the development of supervisors' responsibilities, the acquaintance with educational policies and administrative systems related to Ph.D. students, the observation of academic ethics and norms, the tips of supervisors' physical and mental self-adjustment, etc. All this is conducted in various forms, such as, experts' demonstrations, peer communication and seminars. It will effectively provide the basic, but powerful supervisory skills (both professional and personal) and responsibilities that all supervisors must have in order to understand and connect with the students and drive results, integrating a theory of "becoming a supervisor" into supervisor professional development.

(5) Providing supervisors with a continuum of school mental health services

It is vital to offer students access to a continuum of free, confidential school mental health services to effectively address issues directed to supervisors' work life and safeguard supervisor well-being [61]. It should be pointed out to supervisors that seeking help is not a sign of weakness. After all, teaching is surprisingly one of the most stressful jobs in the United States [62], and perhaps this is even more the case in China. So, supervisors should be encouraged to participate in various amateur activities and communicate with their friends, family members and colleagues in their spare time, which is beneficial to keep a balanced interpersonal relationship and obtain more support from their universities, the community and society. Research also reveals that in addition to the positive benefits of improved teacher job satisfaction, health and well-being, there are documented cost savings and impacts on student outcomes related to having healthy teachers and school staff [63].

5. Limitations

There are three main limitations of this research. Firstly, selection of participants was limited to certain regions, and the sample size was rather small, not indicative of the large population of supervisors across China. Secondly, further analysis should be conducted to cover different levels of stress suffered by participants due to their personal experiences. Thirdly, use of WeChat for data gathering methods tremendously limited the amount of non-verbal data which could have been gathered otherwise, thus limited the probes and follow-up questions concerning the positive or negative influence on the well-being of supervisors following the reforms in higher education.

6. Conclusions

This study brings to light just how stressful the job of doctoral supervision can be, cohering with yet contributing to the emerging and growing body of literature on the psychological health crisis prevalent in higher education. More importantly, this study applied grounded theory to reveal varied factors of academic and interpersonal nature for occupational stress that the participants experienced as doctoral supervisors, and to present a substantive theory that explains the relationship between a variety of related conditions.

Hopefully, this research can provide some insight for certain reforms in respect to doctoral education, arouse concerns from universities and institutions about the psychological health of this special group of people, facilitate the promulgation of related guidelines to alleviate the distressing symptoms among supervisors and promote sustainable development in higher education.

This qualitative study conducted semi-structured interviews to explore the perspectives of the supervisors through inviting them to share their own stories, as it is their perceptions and experiences which are of interest here. Thus, more strategies can be put forward to cope with their stress based on their respective experiences.

Funding: This research was funded by the National Social Science Fund of China, grant number 21BYY091.

Institutional Review Board Statement: The study was conducted in accordance with the Declaration of Helsinki; however, there is no IRB (Institutional Review Board) at Huaqiao University. The participants were introduced by the friends and colleagues of the investigator. At the outset of the interview, each participant was given the form of informed consent. They were informed of the purpose of the study and the voluntary and confidential nature of the interview.

Informed Consent Statement: Informed consent was obtained from all participants involved in the study.

Data Availability Statement: The data presented in this study are not publicly available due to privacy concerns.

Acknowledgments: First, I would like to express my heartfelt gratitude to Jian Wang who gave his valuable and constructive suggestions during the planning and development of this study. His willingness to give his time so generously has been very much appreciated. I would also like to thank all the participants who were interviewed for this study. The article would not have been possible without their valuable time and support.

Conflicts of Interest: The author declares no conflict of interest.

References

1. Adekeye, O.A.; Igbinoba, A.O.; Solarin, M.A.; Ahmadu, F.O.; Owoyomi, O.E. Exploring occupational stress and job involvement of workers in private and public organizations. In Proceedings of the SOCIOINT 2017—4th International Conference on Education, Social Sciences and Humanities, Dubai, United Arab Emirates, 10–12 July 2017.
2. Szymanski, E.M. Disability, job stress, the changing nature of careers, and the career resilience portfolio. *Rehabil. Couns. Bull.* **1999**, *42*, 279–284.
3. Wang, X.Y.; Wang, C.; Wang, J. Towards the contributing factors for stress confronting Chinese Ph.D. students. *Int. J. Qual. Stud. Health Well-Being* **2019**, *14*, 1–11. [CrossRef] [PubMed]
4. Pappa, S.; Elomaa, M.; Perälä-Littunen, S. Sources of stress and scholarly identity: The case of international doctoral students of education in Finland. *High. Educ.* **2020**, *80*, 173–192. [CrossRef]
5. Kinman, G.; Jones, F.; Kinman, R. The Well-being of the UK Academy, 1998–2004. *Qual. High. Educ.* **2006**, *12*, 15–27. [CrossRef]
6. Biron, C.; Brun, J.P.; Ivers, H. Extent and sources of occupational stress in university staff. *Work* **2008**, *30*, 511–522. [CrossRef]
7. Desouky, D.; Allam, H. Occupational stress, anxiety and depression among Egyptian teachers. *J. Epidemiol. Glob. Health* **2017**, *7*, 191–198. [CrossRef] [PubMed]
8. Wang, L.J.; Zhang, D.J. Characteristics of anxiety among primary and middle school teachers: A content-based approach to state anxiety. *Health* **2012**, *4*, 26–31. [CrossRef]
9. Xu, X.H. Jiao shi zhi ye ya li san ci diao cha dui bi yu shi zheng yan jiu [meaning: A contrastive and empirical study of three teacher stress surveys]. *J. Shanghai Educ. Res.* **2017**, *8*, 65–69. [CrossRef]
10. Guo, D.X.; Chu, J.T.; Shi, G.J. Da xue jiao shi cun zai xing jiao lü yan jiu: Zi ben de shi jiao [meaning: A study of the existential anxiety among university teachers: Perspectives from capital]. *Teach. Educ. Res.* **2019**, *31*, 60–67. [CrossRef]
11. Wan, L.; Yang, H.Q. The impacts on overwork of research performance pressure of young college teachers: The intermediary impacts of occupational stress and anxiety. *J. China Univ. Labor Relat.* **2018**, *32*, 47–52.

12. Powell, S.; Green, H. *The Doctorate Worldwide*; Open University Press: Berkshire, UK, 2007.
13. Olson, K.; Clark, C.M. A signature pedagogy in doctoral education: The leader-scholar community. *Educ. Res.* **2009**, *38*, 216–221. [CrossRef]
14. Wisker, G.; Robinson, G. The 'creative-minded supervisor': Gatekeeping and boundary breaking when supervising creative doctorates. In *Postgraduate Supervision: Future Foci for the Knowledge Society*; Fourie-Malherbe, M., Albertyn, R., Aitchison, C., Bitzer, E., Eds.; Sun Press: Stellenbosch, South Africa, 2016; pp. 335–348.
15. Wisker, G.; Robinson, G. Doctoral 'orphans': Nurturing and supporting the success of postgraduates who have lost their supervisors. *High. Educ. Res. Dev.* **2013**, *32*, 300–313. [CrossRef]
16. Bytheway, J. Using grounded theory to explore learners' perspectives of workplace learning [special issue]. *Int. J. Work-Integr. Learn.* **2018**, *19*, 249–259.
17. Corbin, J.; Strauss, A.L. *Basics of Qualitative Research: Techniques and Procedures for Developing Grounded Theory*, 3rd ed.; Sage: Thousand Oaks, CA, USA, 2008; pp. 1–130.
18. Nunes, B.; Martins, T.; Zhou, L.H.; Alajamy, L.; AI-Mamari, S. Contextual sensitivity in grounded theory: The role of pilot studies. *Electron. J. Bus. Res. Methods* **2010**, *8*, 73–84.
19. Van Wijk, E.; Harrison, T. Managing ethical problems in qualitative research involving vulnerable populations using a pilot study. *Int. J. Qual. Methods* **2013**, *12*, 570–586. [CrossRef]
20. Turner, D.W. Qualitative interview design: A practical guide for novice investigators. *Qual. Rep.* **2010**, *15*, 754–760. [CrossRef]
21. Hennink, M.; Hutter, I.; Bailey, A. *Qualitative Research Methods*, 2nd ed.; Sage: London, UK, 2020; pp. 91–114.
22. Jacob, S.A.; Ferguson, S.P. Writing interview protocols and conducting interviews: Tips for students new to the field of qualitative research. *Qual. Rep.* **2012**, *17*, 1–10. [CrossRef]
23. Code of Ethics for Social Workers. Available online: http://xxgk.mca.gov.cn:8011/gdnps/pc/content.jsp?id=14265&mtype=1 (accessed on 31 December 2012).
24. Kathleen, B. Ethical decision-making among hospital social workers. *J. Soc. Work Values Ethics* **2006**, *3*, 61–84. [CrossRef]
25. Norman, L.; Jay, S.; Heidi, H.-L. Triage and ethics: Social workers on the frontline. *J. Hum. Behav. Soc. Environ.* **2008**, *18*, 184–203. [CrossRef]
26. Doody, O.; Noonan, M. Preparing and conducting interviews to collect data. *Nurse Res.* **2013**, *20*, 28–32. [CrossRef] [PubMed]
27. Hand, H. The mentor's tale: A reflexive account of semi-structured interviews. *Nurse Res.* **2003**, *10*, 15–27. [CrossRef] [PubMed]
28. Deamley, C. A reflection on the use of s. *Nurse Res.* **2005**, *13*, 19–28. [CrossRef]
29. Buus, N.; Angel, S.; Traynor, M.; Gonge, H. Psychiatric nursing staff members' reflections on participating in group-based clinical supervision: A semistructured interview study. *Int. J. Ment. Health Nurs.* **2011**, *20*, 95–101. [CrossRef] [PubMed]
30. Dapkus, M.A. A thematic analysis of experience of time. *J. Personal. Soc. Psychol.* **1985**, *49*, 408–419. [CrossRef]
31. Marín, G.; Marín, V.B. *Research with Hispanic Populations*; Sage: Newbury Park, CA, USA, 1991; pp. 82–100.
32. Brislin, R.W. Back-translation for cross-cultural research. *J. Cross-Cult. Psychol.* **1970**, *1*, 185–216. [CrossRef]
33. Morrow, S.L. Quality and trustworthiness in qualitative research in counseling psychology. *J. Couns. Psychol.* **2005**, *52*, 250–260. [CrossRef]
34. Santos, H.P., Jr.; Black, A.M.; Sandelowski, M. Timing of translation in cross-language qualitative research. *Qual. Health Res.* **2015**, *25*, 134–144. [CrossRef] [PubMed]
35. Olson, K. *Essentials of Qualitative Interviewing*; Left Coast Press: Walnut Creek, CA, USA, 2011.
36. Ornstein, A.; Lunenburg, F. *Educational Administration: Concepts and Practices*; Thomson Higher Education: Belmont, CA, USA, 2004; p. 505.
37. Higton, J.; Leonardi, S.; Richards, N.; Choudhoury, A.; Sofroniou, N.; Owen, D. *Teacher Workload Survey 2016: Research Report*; Department for Education: Manchester, UK, 2017.
38. Wingfield, B. How much time does it take to supervise a Ph.D. student? *S. Afr. J. Sci.* **2012**, *108*, 30–32. [CrossRef]
39. The Changing Ph.D.—Turning out Millions of Doctorates. Available online: https://www.universityworldnews.com/post.php?story=20130403121244660 (accessed on 3 April 2013).
40. Elken, M.; Hovdhaugen, E.; Stensaker, B. Global rankings in the Nordic region: Challenging the identity of research-intensive universities? *High. Educ.* **2016**, *72*, 781–795. [CrossRef]
41. Lim, M.A. The building of weak expertise: The work of global university rankers. *High. Educ.* **2018**, *75*, 415–430. [CrossRef]
42. Roberts, G.; Pregitzer, M. Why employees dislike performance appraisals. *Regent Glob. Bus. Rev.* **2007**, *1*, 14–21.
43. Flaniken, F. Performance Appraisal Systems in Higher Education: An Exploration of Christian Institutions. Ph.D. Thesis, University of Central Florida, Orlando, FL, USA, 2009.
44. Murphy, K.R.; Cleveland, J. *Understanding Performance Appraisal: Social, Organizational, and Goal-Based Perspectives*; Sage Publications: Thousand Oaks, CA, USA, 1995.
45. Hermann, K.J.; Wichmann-Hansen, G.; Jensen, T.K. *Quality in the Ph.D. Process: A Survey among Ph.D. Students at Aarhus University*; Aarhus University: Aarhus, Denmark, 2014.
46. Peter Higgs: I Wouldn't Be Productive Enough for Today's Academic System. Available online: https://www.theguardian.com/science/2013/dec/06/peter-higgs-boson-academic-system (accessed on 7 December 2013).
47. Lee, A.; Kamler, B. Bringing pedagogy to doctoral publishing. *Teach. High. Educ.* **2008**, *13*, 511–523. [CrossRef]
48. Moradi, S. Publication should not be a prerequisite to obtaining a PhD. *Nat. Hum. Behav.* **2019**, *3*, 1025. [CrossRef] [PubMed]

49. Controversy Re-Arising as Experts Call for Abolishing Mandatory Requirement that Ph.D. Student Must Publish Two Articles in Journals Indexed in CSSCI before They Graduate. Available online: http://digitalpaper.stdaily.com/http_www.kjrb.com/kjrb/html/2018-01/05/content_385603.htm?div=$-$1 (accessed on 5 January 2018).
50. Bazrafkan, L.; Shokrpour, N.; Yousefi, A.; Yamani, N. Management of stress and anxiety among Ph.D. students during thesis writing: A qualitative study. *Health Care Manag.* **2016**, *35*, 231–240. [CrossRef] [PubMed]
51. Ph.D. Students Face Significant Mental Health Challenges. Available online: http://www.sciencemag.org/careers/2017/04/PhD-students-face-significant-mental-health-challenges (accessed on 4 April 2017).
52. Hunan University Investigates Party Officials for Plagiarism. Available online: https://www.sixthtone.com/news/1004238/hunan-university-investigates-party-officials-for-plagiarism (accessed on 4 July 2019).
53. Prazeres, F. Ph.D. supervisor-student relationship. *J. Adv. Med. Educ. Prof.* **2017**, *5*, 220–221.
54. Maina, M.B.; Bukar, A.M.; Jauro, S.S. Plagiarism: A perspective from a case of a northern Nigerian University. *Int. J. Inf. Res. Rev.* **2014**, *1*, 225–230.
55. PhDs: The Tortuous Truth. Nature. Available online: https://www.nature.com/articles/d41586-019-03459-7 (accessed on 13 November 2019).
56. Kiley, M. Identifying threshold concepts and proposing strategies to support doctoral candidates. *Innov. Educ. Teach. Int.* **2009**, *46*, 293–304. [CrossRef]
57. Kinman, G.; Wray, S. *Higher Stress: A Survey of Stress and Well-Being among Staff in Higher Education*; University and College Union: London, UK, 2013.
58. Lever, N.; Mathis, E.; Mayworm, A. School mental health is not just for students: Why teacher and school staff wellness matters. *Rep. Emot. Behav. Disord. Youth* **2017**, *17*, 6–12.
59. Severinsson, E. Rights and responsibilities in research supervision: Research supervision. *Nurs. Health Sci.* **2015**, *17*, 195–200. [CrossRef] [PubMed]
60. Pratt, K.; Stenning, R. *Managing Staff Appraisal in Schools*; Taylor & Francis: London, UK, 1989; p. 109.
61. Parks, K.M.; Steelman, L.A. Organizational wellness programs: A meta-analysis. *J. Occup. Health Psychol.* **2008**, *13*, 58–68. [CrossRef]
62. Tornio, S. We need to do more for teachers who are exhausted, stressed, and burned out. *Speak Out: The Official Newsletter of the Sachem Central Teachers' Association.* 2018, Volume XXXVI. Available online: https://100iq.com/we-need-to-do-more-for-teachers-who-are-exhausted-stressed-and-burned-out/ (accessed on 9 October 2018).
63. Mattke, S.; Liu, H.S.; Caloyeras, J.; Huang, C.Y.; van Busum, K.R.; Khodyakov, D.; Shier, V. Workplace wellness programs study: Final report. *Rand Health Q.* **2013**, *3*, 7. [PubMed]

International Journal of
*Environmental Research
and Public Health*

Article

How Does Leadership in Safety Management Affect Employees' Safety Performance? A Case Study from Mining Enterprises in China

Shu Zhang [1], **Xinyu Hua** [1], **Ganghai Huang** [2,*] and **Xiuzhi Shi** [1]

[1] School of Resources and Safety Engineering, Central South University, Changsha 410083, China; zhangshu@csu.edu.cn (S.Z.); 205512131@csu.edu.cn (X.H.); baopo@csu.edu.cn (X.S.)
[2] School of Civil Engineering, Central South University, Changsha 410083, China
* Correspondence: huangganghai@csu.edu.cn; Tel.: +86-189-3240-6644

Abstract: Leadership is a necessary element for ensuring workplace safety. Rather little is known about the role of leadership safety behaviours (LSBs) in the mining industry. Using regression analysis and structural equation modelling analysis, this study examined the cause-and-effect relationships between leadership safety behaviours and safety performance. Data were collected by questionnaires from 305 miners in China. Data were analysed using exploratory factor analysis and confirmatory factor analysis, which identified five main dimensions of LSBs: safety management commitment, safety communication with feedback, safety policy, safety incentives, and safety training; the analysis also identified three main dimensions of safety performance: employee's safety compliance, safety participation, and safety accidents. The results showed the overall effects of each LSB variable on safety compliance in descending order as: safety training (0.504), safety incentives (0.480), safety communication with feedback (0.377), safety management commitment (0.281), and safety policy (0.110). The overall effects of each LSB variable on safety participation in descending order were: safety training (0.706), safety incentives (0.496), safety management commitment (0.365), and safety policy (0.247). Furthermore, we found that safety management commitment and safety incentives increased employees' safety behaviours, but this influence was mediated by safety training, safety policy, and safety communication with feedback.

Keywords: safety leadership; safety performance; safety compliance; safety participation

check for
updates

Citation: Zhang, S.; Hua, X.; Huang, G.; Shi, X. How Does Leadership in Safety Management Affect Employees' Safety Performance? A Case Study from Mining Enterprises in China. *Int. J. Environ. Res. Public Health* **2022**, *19*, 6187. https://doi.org/10.3390/ijerph19106187

Academic Editor: Masahito Fushimi

Received: 23 April 2022
Accepted: 14 May 2022
Published: 19 May 2022

Publisher's Note: MDPI stays neutral with regard to jurisdictional claims in published maps and institutional affiliations.

1. Introduction

According to the State Administration of Work Safety in China [1], 5930 accidents and 18,567 deaths occurred in the national mines from 2003 to 2021, second only to the transport industry. These startling facts tell us that the mine safety situation is still grim in China. In 2019, in Inner Mongolia province, accidents while transporting workers down into mines caused 20 deaths and 30 injuries. In 2021, in Shanxi province, a mine flood caused 13 deaths. The same year, in Shandong province, 10 people died, 11 were injured, and 1 was missing in a blast in a return air shaft. Moreover, in addition to China, serious mining accidents have occurred in other countries. In 2021 in Kuzbas, Russia, mining fires occurred that resulted in 52 deaths. In 2022 in Poland, mine collapses resulted in 10 deaths, and a few days later, methane exploded in a coal mine and led to 8 deaths and 7 missing. In addition to the victims' psychological and physical harm, heavy casualties bring families tremendous suffering. From this situation arises two questions: What are the main causes of accidents, and how can safety be improved?

The causes of accidents are complex and closely related to safety management practices [2]. With the growing understanding of the role of safety behaviours in the accident investigation process, people are gradually paying more attention to management deficiencies and leadership behaviours (LSBs). The current research on safety behaviour has made two breakthroughs. On the one hand, it has broadened the range of behaviours beyond employees' working safety behaviours to focus on leaders' management behaviours. For example, Conchie et al. explored factors influencing supervisors' safety leadership in construction [3]. In a survey of about 10,000 workers employed on offshore platforms operating on the Norwegian, Dahl and Olsen indicated that safety leadership affected workers' safety performance through the intervening variable of work climate [4]. Li found that effective mining safety management can prevent accidents [5], and his team analysed the relationship between safety leadership and leadership style [6]. A study focused on Portuguese companies found that safety management behaviours, such as OHS training, largely affected workers' behaviours [7]. On the other hand, the research investigated in depth factors affecting safety behaviour, which included not only the individual's physiological and psychological factors [8] but also organizational factors [9], environmental factors [10,11], and leadership factors [3–5]. This paper mainly studies the LSBs first, then concerns about the consequences of LSBs that influence miners' safety performance.

At present, a large number of studies have shown that LSBs have major impacts on employees' safety behaviour. Generally, leadership safety behaviours (LSBs) were defined as behaviours exhibited by managers when focusing on safety performance [12]. Barling et al. [13] found that supervisors' transformational leadership is positively related to employees' safety behaviours in the hospitality sector, which showed that employees' perceptions of transformational leadership can determine their self-reported safety behaviours. S. Larsson et al. [14] investigated how individual psychological atmosphere affected safety behaviour, again finding that management practices can change employees' safety behaviours. Lu and Yang [15] studied the relationships between safety leadership and employees' self-reported safety behaviours in the container shipping context, and the results suggested that safety incentives, safety policy, and safety concerns positively affected employees' safety participation. Sampling 548 railroad workers, Kath [16] et al. concluded that the communication between workers and superiors has a close relationship with unsafe behaviours. A previous study also found that the core leadership behaviours that positively influenced safety included continuous planning and coordination, role modelling, monitoring work, and proactively correcting deviations [17]. At present, these studies have been conducted in various industries, such as air transportation [18], medical industry [19], mining industry [20], construction [21], electricity industry [22], chemical industry [23], and manufacturing [24].

In a variety of fields, studies on the growing concerns about safety leadership have demonstrated that safety leadership increases organizations' safety performance. Developing and maintaining safety leadership is critical for reducing accidents and promoting safety, and the main purpose of this paper is to explore the relationships between LSBs and employees' safety behaviours in mining enterprises.

1.1. Leadership Safety Behaviours (LSBs)

Many scholars studied aspects of LSBs such as safety management commitment, safety communication with feedback, safety policy, safety incentives, and safety training. Safety management commitment reflects leaders' attitudes toward safety. It plays an important role in organizations' safety programmes [25]. We can understand safety commitment by leaders' safety awareness. Their commitment is reflected in safety rules, regulations, and policies; safety responsibilities; emergency responses; and human, financial, and material resources [26]. Conducting a survey in India, Vinodkumar found that workers at companies certified by OHSAS 18001 and ISO 9001 show more commitment to safety behaviours [27]. Plausible safety commitment from leaders may influence workers' psychological capital, in turn improving their safety performance [28]. Good communication and feedback will help employees gain experience and promote safety awareness effectively in the event of accidents. Regular communication between managers and employees can also improve the safety atmosphere in workplaces. Some studies showed that safety communication with feedback significantly affected organizational safety [16,28]. Konstantin et al. also indicated that safety communication greatly affected workers' safety behaviours [29]. Following a literature review, Cohen concluded that abundant safety communication raises workers' safety awareness [30]. Carrillo and Simon [31] and Barling et al. [13] considered safety communication with feedback one dimension of LSBs. Other researchers, such as Hofmann and Morgeson [32], Mearns et al. [33], Probst [34], Zuo [35], and Oswald and Lingard [36], also confirmed that safety communication was negatively related to occupational accidents. Safety policy consists of safety regulations and procedures and represents the safety standards of enterprises. Reasonable safety policies help employees clearly understand safety requirements and thus improve employees' safety behaviours [15,27,37,38]. Safety incentives also reflect one dimension of LSBs. Appropriate safety incentives can motivate and strengthen employees' safety behaviours [15,39,40]. As such, establishing a fair evaluation and reward system can effectively avoid unsafe behaviours. For enterprise managers and ordinary employees, safety management effects are inseparable from those of safety education and training. Safety training can not only improve employees' skills in risk identification and crisis management but also improve their safety knowledge and awareness. Improper safety training negatively affects employees' safety behaviours [41,42].

1.2. Enterprise Safety Performance (ESP)

The traditional assessment of safety performance has been mainly based on accidents, near-misses, injuries, diseases, and other objective results. Currently, safety performance also includes safety behaviours and accident rates. Neal et al. [43] considered safety compliance and safety participation components of safety performance. Through exploring the influencing factors of safety performance, Abbas [44] found that the relationships among managers, harmonious environments, and work flexibility significantly influenced safety performance. From organizational levels, Claudio et al. [45] identified two elements and two important influencing factors for safety performance: safety participation, safety norms, safety habits, and safety compliance. The lean philosophy developed to improve safety management and safety performance [46], and Cordeiro et al. identified that lean tools—5S, visual management, and OPL—are important for improving safety conditions and promoting a safety culture as part of safety management [47].

1.3. Research Hypothesis

To summarize the literature findings, a considerable number of domestic and foreign scholars have studied LSBs and obtained many valuable research results. Prior studies generally agreed that LSBs included safety management commitment, safety communication with feedback, safety policy, safety incentives, and other dimensions; most of the studies used surveys. At the same time, many scholars studied how LSBs affect employees' safety behaviours. Whether a safety culture or climate has successfully been established has a vital link with leaders' attitudes, behaviours, and decision-making. Leaders' words greatly influence employees' views and ideas, and we argue that LSBs include some dimensions of safety culture or climate.

Although management and production are different in the mining industry than in other industries, there are some similarities such as the high risks of construction, managers' high safety responsibilities in marine transportation, and the need for specific safety measures such as in machinery and chemical industries. Based on the existing research, we aimed to study five dimensions of LSBs: safety management commitment (SMC), safety communication with feedback (SCF), safety policy (SP), safety incentives (SI), and safety training (ST), and three dimensions of enterprise safety performance, safety compliance, safety participation, and safety accidents.

Based on the literature, we proposed the following hypotheses:

Hypothesis 1 (H1). *Among LSBs, safety training (1), safety management commitment (2), safety incentives (3), safety policy (4), and safety communication with feedback (5) will affect safety compliance directly, significantly, and positively.*

Hypothesis 2 (H2). *Among LSBs, safety training (1), safety management commitment (2), safety incentives (3), safety policy (4), and safety communication with feedback (5) will affect safety participation directly, significantly, and positively.*

Hypothesis 3 (H3). *Among LSBs, safety training (1), safety management commitment (2), safety incentives (3), safety policy (4), and safety communication with feedback (5) will affect safety accidents directly, significantly, and positively.*

2. Materials and Methods

2.1. Sample

For this study, we conducted a questionnaire survey in lead-zinc mines in China. The survey was sent to 450 employees in middle, primary, and workshop. A total of 450 questionnaires were returned, for a response rate of 100%. A total of 145 invalid questionnaires were collected, and 305 valid questionnaires were obtained, an effective rate of 68%. Thus, the number of samples satisfied statistical requirements.

2.2. Measures

2.2.1. Independent Variables

The employees' perceptions about the five dimensions of LSBs were measured with a questionnaire based on a review of related literature and theories [31,39,48,49]. The questionnaire contained questions covering safety management commitment (9 items), safety communication with feedback (5 items), and safety training (6 items) designed with reference to previous studies [27,28,37,43,50–52]. In addition, it contained questions covering safety policy (4 items) and safety incentives (6 items) which referred to Cooper [39], Carrillo and Simon [31], O'Dea and Flin [48], Wu et al. [49], and Lu and Yang [15]. This portion of the questionnaire had a total of 30 items (see Appendix A). In this research, respondents rated items on five-point Likert scales (anchored by 1 = "strongly disagree" and "5 = strongly agree").

2.2.2. Dependent Variables

ESP was measured on three dimensions: safety compliance, safety participation, and safety accidents. In previous studies, self-reported safety behaviour referred to safety compliance and safety participation. Based on the literature review, we adapted safety compliance (4 items) and safety participation (3 items) from Borman and Motowidlo [53], Neal et al. [43], and so on. We added one item to the dimension of safety compliance: "In order to complete more work to get more piece-rate income or measurement of income, I may ignore safety". This question acknowledged the actual mining situation in China. In addition, safety incidents were measured with the item "In the past three years, I have been in an accident", taken from Leung et al. [42]. In total, the safety performance questionnaire consisted of 9 items (see Appendix B). Respondents assessed items using five-point Likert scale.

2.2.3. Control Variables

In addition to LSBs and safety performance, the questionnaire included some control variables. Respondents' sex, age, education level, work experience, job position, and frequency of safety training were the control variables. Education level and job position reflected the respondents' understanding of safety leadership and safety behaviours. The frequency of safety training indicated whether respondents had abundant safety knowledge, which would affect their understanding of safety behaviours.

2.3. Data Analysis

Firstly, exploratory factor analysis was used to confirm the dimensions of LSBs and safety performance from the questionnaire. Then, confirmatory factor analysis was used to examine the reliability and validity of the questionnaires on LSBs and safety performance. Next, regression analysis was conducted to explore LSBs' influence on safety performance. Last, SEM analysis was conducted to study LSBs' influencing mechanism on safety performance.

2.3.1. Questionnaire of LSBs

Exploratory Factor Analysis

Based on 50% of the sample, SPSS16.0 was used for exploratory factor analysis. Principal component analysis with varimax rotation was employed to identify the dimensions of LSBs. The results showed that only variables with a factor loading greater than 0.50 were extracted; the retained items' factor loadings ranged from 0.527 to 0.907. The total explained variance of the common LSB factors amounted to 70.897%, and Cronbach's $\alpha > 0.60$, which indicated that the five dimensions of LSBs were persuasive. Through exploratory factor analysis, a total of 13 items were deleted: X4, X5, X6, X7, X11, X12, X14, X15, X18, X21, X23, X24, and X28. Thus, a five-dimensional structure was formulated. Then, factor analysis was conducted to confirm the reliability and stability of the five dimensions of LSBs.

Factor 1 was called safety training and accounted for 59.66% of the total variance. Factor 2 was called safety management commitment and accounted for 18.335% of the total variance. Factor 3 was designated safety incentives and accounted for 12.697% of the total variance. Factor 4 was called safety policy and accounted for 9.307% of the total variance. Factor 5 was designated safety communication with feedback and accounted for 9.119% of the total variance.

Confirmatory Factor Analysis

Structural equation modelling (SEM) confirmatory factor analysis was used to study the reliability and validity of the LSB questionnaire [54].

(1) Fit test of the structural model

Following the results of exploratory factor analysis, two models for hypothesis testing were proposed: a single-factor model and a five-factor model (Figures 1 and 2).

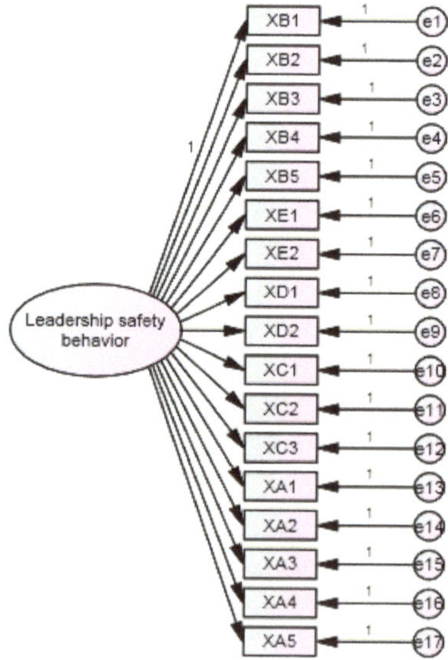

Figure 1. The single-factor structural model.

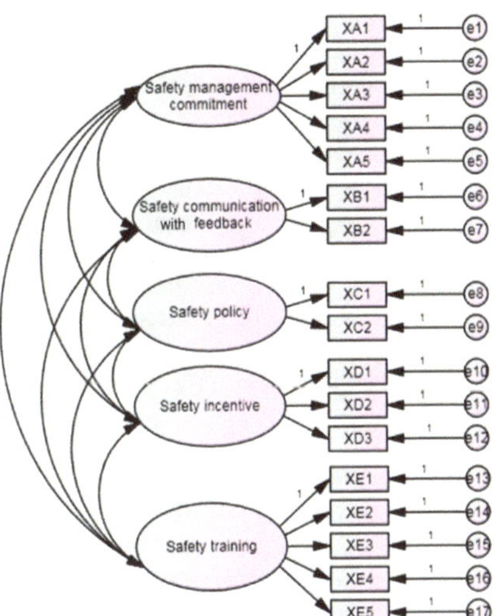

Figure 2. The five-factor structural model.

The results showed a chi-square degrees of freedom ratio ($\chi^2/\mathrm{df} = 1.205$) of less than 2 and an RMSEA (0.038) of less than 0.08 that indicated that the five-factor structure model fit better. At the same time, although the AGFI of the five-factor model did not meet 0.90, the model also did not have a negative error variable. All factor loadings ranged from 0.50 to 0.95, the standard error was fair, all parameters were significant, and the standardized residuals with absolute value were less than 2.58. Overall, the five-factor model fit well and had good construct validity. Thus, the five-factor model could be accepted.

(2)   Reliability

The reliability of the structural model could be assessed by Cronbach's α and construct reliability. Generally, Cronbach's α more than 0.60 yields high confidence of a model. Construct reliability more than 0.6 indicates ideal intrinsic quality of the model and that the measurement tool is stable. The results showed that the Cronbach's α of the latent variables ranged from 0.616 to 0.869 (>0.6), and the construct reliability ranged from 0.656 to 0.873 (>0.6). The measurements of each latent variable had good internal consistency.

(3)   Validity

Convergent validity can be confirmed by *t*-values which are all statistically significant on the factor loadings. In the AMOS text output file, the *t*-value is the critical ratio (C.R.). The larger the factor loadings or coefficients, the stronger the observed variables' representation of their specified latent variables. Each item exceeds the critical ratio at the 0.05 level of significance (*p*). Thus, all observed variables were significantly related to their specified latent variables. Item reliability refers to the R^2 value, which can be used to estimate the reliability of a particular observed variable (item). $R^2 > 0.50$ provides evidence for acceptable reliability. Although, some items' R^2 values were less than 0.50, generally speaking, the relationships between the latent variables and the observed variables were reasonable, and this model had good convergent validity.

Discriminant validity can be measured by comparing the goodness of fit before and after the merger of two factors. The results showed the discriminant validity between 10 paired factors. The difference of chi-square value between the restricted model and the unrestricted model reached the 0.05 level of significance. The results provided evidence of good discriminant validity.

In addition, the variance extracted can also measure construct reliability. High variance extracted values occur when the observed variables are truly representative of the latent variables. The results showed that the variance extracted values are: 0.476, 0.511, 0.576, 0.583, and 0.626. Overall, most of the latent variables had a variance extracted value that was higher than the recommended level of 50%. This indicated that the overall goodness-of-fit results supported the proposed model.

Based on the results above, the five-factor LSB model had high reliability and validity and could be used for further research.

2.3.2. Questionnaire of Safety Performance

Using AMOS18.0, confirmatory factor analysis was used to examine the reliability and validity of the safety performance questionnaire. The calculation methods were similar to the foregoing.

The results showed that the chi-square degrees of freedom ratio ($\chi^2/df = 1.494$) was less than 2 and the RMSEA (0.059) was less than 0.08, indicating that the three-factor model of safety performance fit well. Cronbach's α of the latent variables were 0.790 and 0.806 (>0.6), and the construct reliability values were 0.765 and 0.806 (>0.5). Hence, the measurements of each latent variable had good internal consistency. For convergent validity, the standardized factor loadings of all items reached the level of significance, which indicated that the questionnaire had high convergent validity.

Based on the results above, the three-factor model of safety performance, with high reliability and validity, could be used for further research.

2.3.3. Regression Analysis

Regression analysis was used to study the relationships between LSBs and safety performance, including simple linear regression analysis and stepwise multiple regression analysis [55].

First, simple linear regression analysis was conducted with the five dimensions of LSBs as independent variables and safety performance as the dependent variable. The results showed that safety training, safety management commitment, safety incentives, safety policy, and safety communication with feedback significantly positively predicted safety performance, accounting for, respectively, 76.5%, 62.5%, 49.6%, 49.9%, and 25.1% of the variance in safety performance (see Table 1). This indicates that the better the safety training, the better the employees' safety performance, which is the same as safety management commitment, safety incentives, safety policy, and safety communication with feedback.

Table 1. The simple linear regression analysis coefficients of the five dimensions of LSBs.

Model	Independent Variable	R	R^2	ΔF	F	Beta	t	Sig.
1	ST	0.875	0.765	455.192 ***	455.192 ***	0.875	21.335 ***	0.000
1	SMC	0.790	0.625	233.100 ***	233.100 ***	0.790	15.268 ***	0.000
1	SI	0.704	0.496	137.810 ***	137.810 ***	0.704	11.739 ***	0.000
1	SP	0.706	0.499	139.318 ***	139.318 ***	0.706	11.803 ***	0.000
1	SCF	0.501	0.251	46.801 ***	46.801 ***	0.501	6.841 ***	0.000

*** At the 0.001 level (three tailed), the correlation is significant.

Then, stepwise multiple regression analysis was conducted with the five dimensions of LSBs as independent variables and safety performance as the dependent variable. Three predictor variables, safety training, safety communication with feedback, and safety policy, significantly predicted safety performance ($R = 0.765$, $R^2 = 0.585$, $F = 64.879$ ***). That is, a total of three predictor variables effectively explained 58.5% of the variance in safety performance. Safety training had the largest predictive power with 52.3% of the variance (see Table 2). The standardized regression coefficients (Beta) of the three predictor variables were positive, which demonstrated that the variables positively affected safety performance. However, separate regression analysis revealed relationships of safety management commitment with safety performance and safety incentives with safety performance, meaning that these two variables were significant positive predictors of safety performance. In turn, this finding meant that when safety training, safety policy, and safety communication with feedback were added to the relationships between safety management commitment, safety incentives and safety performance, safety management commitment and safety incentives showed significantly lower predictive power for safety performance, which indicated that the first three factors likely played significant mediating effect.

Table 2. The stepwise multiple regression analysis coefficients of the five dimensions of LSBs.

Input Variable Order	R	R^2	F	ΔF	B	Beta	TOL	VIF	Eigen-Values	CI
ST	0.723	0.523	153.590 ***	153.590 ***	0.800	0.554	0.625	1.599	0.071	7.422
SCF	0.754	0.569	91.749 ***	14.785 ***	0.599	0.231	0.862	1.160	0.017	15.155
SP	0.765	0.585	64.879 ***	5.370 *	0.575	0.152	0.695	1.439	0.013	17.491

*** At the 0.001 level (three tailed), the correlation is significant. * At the 0.05 level (three tailed), the correlation is significant.

2.3.4. SEM Analysis

Regression analysis indicated the partial mediation of safety training, safety policy, and safety communication with feedback. Therefore, considering safety training, safety policy, and safety communication with feedback as mediator variables, the effects of LSBs on safety performance were further studied by SEM [56]. The following hypotheses were proposed:

Hypothesis 4 (H4). *Safety training (1), safety policy (2), and safety communication with feedback (3) will mediate the relationship between safety management commitment and employee safety compliance.*

Hypothesis 5 (H5). *Safety training (1), safety policy (2), and safety communication with feedback (3) will mediate the relationship between safety management commitment and employee safety participation.*

Hypothesis 6 (H6). *Safety training (1), safety policy (2), and safety communication with feedback (3) will mediate the relationship between safety management commitment and safety accident.*

Hypothesis 7 (H7). *Safety training (1), safety policy (2), and safety communication with feedback (3) will mediate the relationship between safety incentive and employee safety compliance.*

Hypothesis 8 (H8). *Safety training (1), Safety policy (2), and Safety communication with feedback (3) will mediate the relationship between safety incentive and employee safety participation.*

Hypothesis 9 (H9). *Safety training (1), Safety policy (2), and Safety communication with feedback (3) will mediate the relationship between safety incentive and safety accident.*

To test the hypotheses above, a structural model was established through SEM (Figure 3). There were eight latent variables in the structural model: safety management commitment, safety communication with feedback, safety policy, safety incentives, safety training, safety compliance, safety participation, and safety accidents. The first model fit better ($\chi^2/\mathrm{df} = 1.102 < 2$, RMSEA = 0.027 < 0.08, see Figure 3) and was revised by deleting the negative paths (Figure 4).

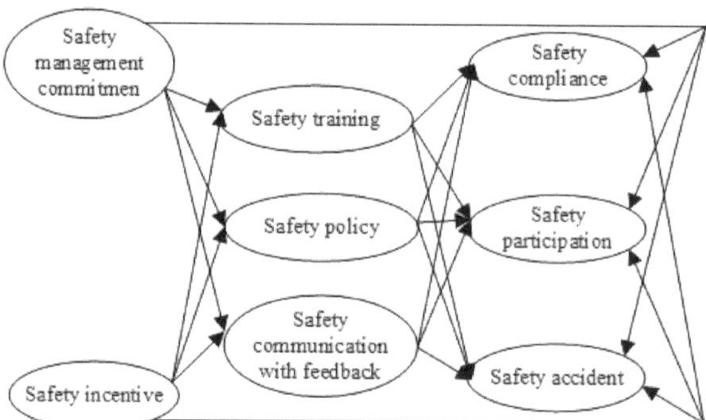

Figure 3. The partial mediation model of leadership behaviours' effects on safety performance (the first model).

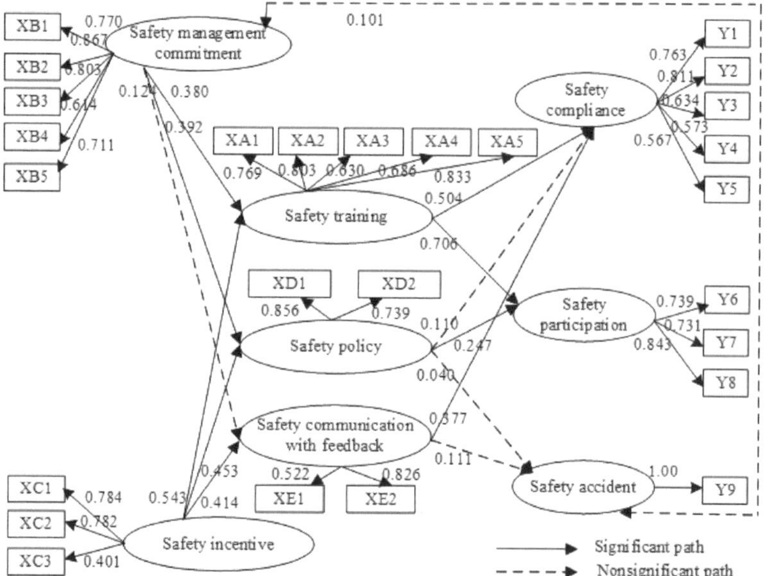

Figure 4. The revised partial mediator model of leadership behaviours' effect on safety performance (the second model).

In the revised model (χ^2/df = 1.126 < 2, RMSEA = 0.030 < 0.08), safety management commitment had no significant effect on safety compliance, safety participation, or safety accidents. However, when the three mediators, safety training, safety policy, and safety communication with feedback, were added, safety management commitment and safety incentives significantly influenced safety compliance and safety participation (Figure 4). This result confirmed the mediation hypotheses above. Thus, the fifth model was selected. Next, the effects of the five dimensions of LSBs on the three dimensions of safety performance will be further analysed.

3. Results

Based on the results of SEM analysis, we illustrated the effect of LSBs on safety performance through the standardized path coefficients; the direct, indirect, and overall effects; and the mediator effects.

3.1. Standardized Path Coefficient

If C.R. > 1.96, the path is significant (see Table 3). The results showed that safety management commitment (SMC) significantly and positively affects safety training (ST, C.R. = 4.314) and safety policy (SP, C.R. = 4.001), which is in line with the hypothesis. That is, leaders' convincing safety management commitment and emphasis on safety at work improves safety training and safety policy. In addition, safety management commitment does not significantly affect safety communication with feedback (SCF, C.R. = 1.066).

Table 3. The standardized path coefficients of the second model.

Internal Latent Variable		Exogenous Latent Variable	Regression Weights	C.R.	*p*	Standardized Regression Weights
SP	<—	SMC	0.438	4.001	***	0.392
SCF	<—	SMC	0.140	1.066	0.286	0.124
ST	<—	SI	0.567	5.456	***	0.543
SP	<—	SI	0.441	4.488	***	0.453
SCF	<—	SI	0.408	2.821	0.005	0.414
ST	<—	SMC	0.455	4.314	***	0.380
Safety compliance	<—	ST	0.630	4.952	***	0.504
Safety participation	<—	ST	0.607	6.575	***	0.706
Safety compliance	<—	SP	0.147	1.265	0.206	0.110
Safety participation	<—	SP	0.227	3.051	0.002	0.247
Safety accident	<—	SP	0.046	0.364	0.715	0.040
Safety compliance	<—	SCF	0.499	3.769	***	0.377
Safety accident	<—	SMC	0.129	0.950	0.342	0.101
Safety accident	<—	SCF	0.125	1.210	0.226	0.111

*** At the 0.001 level (two tailed), the correlation is significant.

Safety incentives (SI) significantly and positively affect safety training (ST, C.R. = 5.456), safety policy (SP, C.R. = 4.488), and safety communication with feedback (SCF, C.R. = 2.821), consistent with the hypothesis. The finding indicates that proper and compelling safety incentives can improve the effects of safety training and safety policy and improve safety communication.

Safety training (ST) significantly and positively affects safety compliance (C.R. = 4.952) and safety participation (C.R. = 6.575): Regular and effective safety training improves employees' safety compliance and increases their participation in safety activities.

Safety communication with feedback (SCF) significantly and positively affects safety compliance (C.R. = 3.769). Forming a good safety communication cycle among employees and leaders may enhance employees' incentives to comply with safety regulations, rules, and operations.

3.2. Direct, Indirect, and Overall Effects

Table 4 showed the various effects of LSBs on safety performance. The following conclusions can be drawn:

(1) On all dimensions of LSBs, safety training has the greatest effect on employees' safety compliance and safety participation.
(2) No dimensions of LSBs had significant effects on safety accidents.
(3) Safety management commitment has no direct, significant, positive impacts on employee safety compliance, safety participation, or safety accidents. Thus, H1(2), H2(2), and H3(2) are not supported. Additionally, safety management commitment has indirect, significant, and positive effects on employees' safety compliance and safety participation.
(4) Safety incentives have no direct, significant, positive effects on employees' safety compliance, safety participation, or safety accidents, which rejects H1(3), H2(3), and H3(3). However, safety incentives have indirect, significant, and positive effects on employees' safety compliance and safety participation.

(5) Safety training and safety communication with feedback have direct, significant, and positive effects on employees' safety compliance. Safety training and safety policy have direct, significant, and positive effects on employees' safety participation. Thus, H1(1), H1(5), and H2(1) are supported.

Table 4. The direct, indirect, and overall effects of LSBs on safety performance.

Variable	Safety Compliance			Safety Participation			Safety Accident		
	Direct Effect	Indirect Effect	Overall Effect	Direct Effect	Indirect Effect	Overall Effect	Direct Effect	Indirect Effect	Overall Effect
Mediator variable									
ST	0.504 *	—	0.504 *	0.706 *	—	0.706 *	—	—	—
SP	0.110	—	0.110	0.247 *	—	0.247 *	0.040	—	0.040
SCF	0.377 *	—	0.377 *	—	—	—	0.111	—	0.111
Exogenous variable									
SMC	—	0.281 *	0.281 *	—	0.365 *	0.365 *	0.101	0.029	0.131
SI	—	0.480 *	0.480 *	—	0.496 *	0.496 *	—	0.064	0.064

Note: "—" means no effect. * At the 0.05 level (one tailed), the correlation is significant.

3.3. Mediator Effects

According to Figure 4 and Tables 3 and 4, the mediator effect of the model was analysed as follows:

(1) In this study, the hypothesis that safety communication with feedback will mediate the relationships between safety management commitment and safety compliance, safety participation, and safety accident is not confirmed.

(2) Safety training fully mediates the effects of safety management commitment on safety compliance and safety participation. Via safety training ($\beta = 0.380$, $p = 0.000$), safety management commitment positively and significantly affects safety compliance ($\beta = 0.504$, $p = 0.000$) and safety participation ($\beta = 0.70$, $p = 0.000$). Thus, H4(1) and H5(1) are supported, and H6(1) is not supported.

(3) Safety policy fully mediates the effect of safety management commitment on safety participation. Via safety policy ($\beta = 0.392$, $p = 0.000$), safety management commitment has a significant and positive effect on safety participation ($\beta = 0.706$, $p = 0.000$). Thus, the results accept H5(2) and reject H4(2) and H6(2).

(4) Safety communication with feedback does not mediate the relationships between safety management commitment and safety compliance, safety participation, or safety accidents. Thus, H4(3), H5(3), and H6(3) are not supported.

(5) Safety training fully mediates the effects of safety incentives on safety compliance and safety participation. Via safety training ($\beta = 0.543$, $p = 0.000$), safety incentives have significant and positive effects on safety compliance ($\beta = 0.504$, $p = 0.000$) and safety participation ($\beta = 0.706$, $p = 0.000$). The results accept H7(1) and H8(1) and reject H9(1).

(6) Safety policy fully mediates the effect of safety incentives on safety participation. Via safety policy ($\beta = 0.453$, $p = 0.000$), safety incentives have a significant and positive effect on safety participation ($\beta = 0.706$, $p = 0.000$). Thus, H8(2) is supported, and H7(2) and H9(2) are not supported.

(7) Safety communication with feedback fully mediates the effects of safety incentives on safety compliance. Via safety communication with feedback ($\beta = 0.414$, $p = 0.005$), safety incentives have a significant and positive effect on safety compliance ($\beta = 0.504$, $p = 0.000$). Thus, H7(3) is accepted, and H8(3) and H9(3) are not accepted.

3.4. General Conclusions

(1) Leadership safety management commitment, safety incentives, safety training, and safety communication with feedback have significant positive effects on employees' safety compliance.

(2)　Leadership safety management commitment, safety incentives, safety training, and safety policy have significant positive effects on employees' safety participation.

(3)　Leadership safety management commitment, safety incentives, safety training, safety policy, and safety communication with feedback have no significant positive effects on safety accidents.

(4)　The overall effects of each variable on safety compliance in descending order are: safety training (0.504 *), safety incentives (0.480 *), safety communication with feedback (0.377 *), safety management commitment (0.281 *), and safety policy (0.110).

(5)　The overall effects of each variable on safety participation in descending order are: safety training (0.706 *), safety incentives (0.496 *), safety management commitment (0.365 *), and safety policy (0.247 *).

(6)　The overall effects of each variable on safety accidents in descending order are: safety management commitment (0.131), safety communication with feedback (0.111), safety incentives (0.064), and safety policy (0.040).

(7)　The path of effects on employees' safety behaviour is shown in Figure 5.

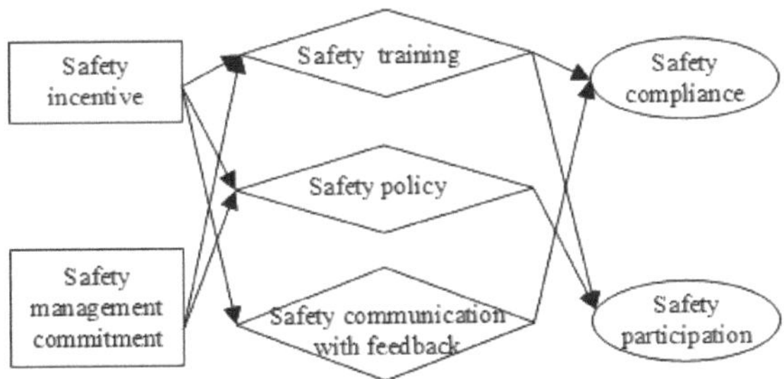

Figure 5. The path of effects on employee safety behaviour.

4. Discussion

This study analysed the overall effects of the five dimensions of LSBs on safety performance. The results show that leaders' safety attitudes and safety behaviours directly affect employees' safety behaviours and play an important role in safety performance. Meanwhile, the results indicate that safety training, safety policy, and safety communication with feedback are significant mediators in predicting safety performance.

4.1. Implications of This Study's Findings

According to the results, the affecting factors included leaders' safety training, safety management commitment, safety incentives, safety policy, and safety communication with feedback, which predicted safety performance significantly and positively. However, considering the comprehensive effects on safety performance, safety management commitment and safety incentives were mediated by safety training, safety policy, and safety communication with feedback. According to the affecting path and overall effects, the conclusions are summarized as follows.

4.1.1. The Factors That Affect Employees' Safety Compliance

In keeping with the findings by Leung et al. [42], leaders' safety training and safety communication with feedback have direct, significant, and positive effects on employees' safety compliance. Effective safety training can tell employees what regulations and rules they should comply with, and good communication helps employees increase safety awareness and avoid unsafe behaviours. If leaders provide a variety of safety training opportunities and encourage employees to participate, employees' safety compliance will be promoted. Leaders could establish smooth communication regarding risks and clear safety goals, which will also effectively promote employees' safety compliance [16,29,57,58].

Safety management commitment has an indirect, significant, and positive effect on employees' safety compliance by the mediator of safety training. Positive and convincing safety management commitment can increase employees' initiative to join safety training and safety activities. Through regular safety training, employees can understand safety regulations and rules more comprehensively and upgrade their skills to avoid unsafe operations, which ultimately strengthens their safety compliance. In addition, when employees behave dangerously, leaders should take measures timely, and enhancing safety training is one of the most important measures in safety management [59]. When accidents happen, leaders need to stress hazard identification skills, which mostly come from safety training [60]. Overall, leaders' safety management commitments, such as attitudes towards complying with safety rules and regulations, and dealing with safety issues, are embodied in safety trainings, which ultimately affect employees' safety compliance.

Safety incentives have an indirect, significant, and positive effect on employees' safety compliance mediated by safety training and safety communication with feedback, which is different from Lu and Yang [15]. The difference may be associated with the conditions of mining in China. Although leaders reward employees who do well in safety behaviour, the evaluation standard is zero accidents and the realization of production targets. This stance does not reward or encourage employees' safe work; in contrast, it may promote them to hide hazards in order to obtain more rewards [61]. Thus, safety incentives have no direct effect on safety compliance in this study.

Secondly, according to incentive theory, expectation theory, and goal-setting theory, with appropriate safety incentives, employees hope to improve their operation skills, safety consciousness, and safety performance by taking part in various trainings. Given this, leaders should conduct effective safety training and good safety communication to meet employees' safety needs, which can increase employees' desire to act more safely and their safety compliance. In addition, in the workplace, if leaders timely praise employees for their risk identification, rather than habitually criticizing employees for their mistakes, such positive reinforcement can promote employees' safety work advantageously. Employees expect to communicate risks and accidents with leaders, and communications about safety aims can motivate employees' safety performance. Overall, these results confirm incentive theory to a certain degree.

Different from previous studies [15,33,37,50,62], the results showed that safety policy had a positive effect on safety compliance which was not significant. Safety policy includes two aspects, emphasis on site safety and establishment of safety responsibility systems [63]. Leaders always focus on the safety status of underground operations. Although leaders inspect workplaces every day, this inspection temporarily promotes safe work. As for employees, this is a passive promotion because employees' safety awareness is the root cause for safety work [64]. Secondly, even when leaders establish a sound system of safety responsibility, how to effectively implement safety responsibility is more important. To encourage employees' responsibility, it is more crucial to reward employees in accordance with safety requirements [17]. This positive reinforcement has more significance for safety compliance. Thus, the effect of safety policy on safety compliance is not significant in our study.

4.1.2. The Factors That Affect Employee Participation

Safety training and safety policy of leadership have direct, significant, and positive effects on employees' safety participation. Safety training can not only improve employees' safety awareness, integration in the enterprise, and understanding of safety attitudes but also encourage employees to comply with mutual responsibility systems. The same as previous studies [15,62], for safety policy, leaders should establish safety responsibility systems and make clear safety accountability at all levels, which can further strengthen employees' responsibility for their own behaviours. The responsibility employees take will increase their attention to daily safety training and increase their initiative to participate in safety activities.

Safety management commitment has an indirect, significant, and positive effect on employees' safety participation mediated by safety training and safety policy. Safety management commitment has no direct and positive effect on safety participation, which is different from Lu [15]. In the survey of enterprises, even when leaders have clear safety attitudes and timely correct employees' unsafe behaviours, employees have little chance to participate in safety management. Thus, these activities have little direct impact on employees' participation in safety activities and meetings or on promoting workplace safety. In contrast, according to the results above, if leaders maintain emphasis on safety, deal with accidents timely, and take the lead in observing safety regulations, their efforts will accentuate the effects of conducting safety training and safety policy. Finally, employees' safety participation will improve.

Safety incentives have indirect, significant, and positive effects on employees' safety participation mediated by safety training and safety policy. Although leaders reward those who set examples of safety, the rewards are based on indicators of accidents rather than on employees' participation in safety activities. As such, safety incentives affect safety participation indirectly. Proper and adequate safety incentives for employees can determine the effects of safety management activities, such as carrying out safety training and enacting safety policy. These activities can enhance employees' safety awareness and encourage employees to actively participate in safety management [65]. In emphasizing site safety, it is also important to encourage employees to ensure the safety of colleagues.

Previous studies showed that effective safety communication with feedback can significantly promote safety participation [16,57]. However, safety communication with feedback has a positive but nonsignificant effect on employees' safety participation. The causes of this phenomenon are mainly features inherent to the mining industry. Miners, managers, and leaders occasionally communicate with each other, so the influence on safety participation exists, but it is not significant, which may be different from other studies.

4.1.3. The Factors That Affect Safety Accidents

Safety management commitment, safety policy, and safety communication with feedback have positive effects on safety accidents (fewer accidents are indicated by higher scores), but not significantly. The influence from safety management commitment is the greatest. This result is consistent with previous research [3,32–34,66,67]. Safety management commitment is perceived by employees as leaders' attitudes and methods of dealing with safety problems. Leaders' positive behaviours, such as handling risks seriously and timely, can possibly reduce accidents. In safety policy, if leaders emphasize practical workplace safety and set up effective safety responsibility systems, the efforts will decrease accidents [68]. Safety communication with feedback is a beforehand control measure perceived by employees [69]. Whether employees communicate hazards and risk information before accidents, whether leaders encourage employees to report dangers, and whether leaders propose clear safety goals and requirements can all predict safety accidents. Finally, different from previous research, safety incentives have no influence on safety accidents. The main reason is that valid safety incentives are lacking in mining enterprises.

4.2. Limitations and Future Research

Even after the exploratory factor analysis, the LSB questionnaire items mainly referred to existing scales, but there is a lack of research on whether these items conform to the situation of the mining industry in China. Therefore, it is hoped that future surveys will better reflect the actual circumstances of mining in China.

Owing to the time and cost, the study only surveyed a few representative mining enterprises. Although the conclusions are more targeted, the results have limitations in overall representativeness. In future studies, we can survey more enterprises.

To date, the studies on LSBs and employees' safety behaviours have almost all adopted self-report measures, and this study is no exception. As a result, the findings are very dependent on the integrity of the respondents, and some respondents may answer the questions casually and unreasonably. Therefore, in order to understand the behaviours of mining enterprises and their affecting factors more accurately, other methods should be used such as interviews and behavioural experiments.

5. Conclusions

Regression analysis and structural equation modelling analysis were used to examine the cause-and-effect relationships between leadership safety behaviours and safety performance. Data collected from a survey of 305 miners were analysed by exploratory factor analysis and confirmatory factor analysis. The following conclusions may be drawn from the results:

(1)　Safety management commitment, safety incentives, safety training, and safety communication with feedback have significant positive effects on employees' safety compliance.

(2)　Safety management commitment, safety incentives, safety training, and safety policy have significant positive effects on employees' safety participation.

(3)　The overall effects of each variable on safety compliance in descending order are: safety training (0.504 *), safety incentives (0.480 *), safety communication with feedback (0.377 *), safety management commitment (0.281 *), and safety policy (0.110).

(4)　The overall effects of each variable on safety participation in descending order are: safety training (0.706 *), safety incentives (0.496 *), safety management commitment (0.365 *), and safety policy (0.247 *).

(5)　The overall effects of each variable on safety accidents in descending order are: safety management commitment (0.131), safety communication with feedback (0.111), safety incentives (0.064), and safety policy (0.040).

(6)　Safety management commitment and safety incentives predicted employees' improved safety behaviours, but this influence was mediated by safety training, safety policy, and safety communication with feedback.

In conclusion, we established a partial mediation model of leadership behaviours' effects on safety performance, and we analysed the effects of leadership behaviours' five dimensions on safety performance's three dimensions. The results provide greater insights into the value of increasing safety performance in mining enterprises. Understanding how leadership behaviours affect employees' safety performance will help leaders to effectively improve safety management behaviours.

Author Contributions: Conceptualization, S.Z.; methodology, S.Z.; software, S.Z.; validation, S.Z., X.H. and X.S.; formal analysis, G.H.; investigation, S.Z.; resources, S.Z.; data curation, X.H.; writing original draft preparation, X.H.; writing, review and editing, X.H.; visualization, S.Z.; supervision, S.Z.; project administration, S.Z.; funding acquisition, S.Z. All authors have read and agreed to the published version of the manuscript.

Funding: This research was funded by the National Natural Science Foundation of China, grant number 51604300.

Institutional Review Board Statement: Ethical review and approval were waived for this study because our research did not entail clinical trials of human or animals, and the questionnaire was formed to investigate employees' opinions on safety management and safety performance, which did not encompass ethical issues.

Informed Consent Statement: Informed consent was obtained from all subjects involved in the study.

Data Availability Statement: Not applicable.

Acknowledgments: The authors want to express thanks to everyone who supported this study.

Conflicts of Interest: The authors declare no conflict of interest.

Appendix A

Table A1. Items on the leadership safety behaviours questionnaire.

Management commitment

1	Leadership attaches great importance to safety issues
2	Safety rules and procedures are strictly followed by leaders
3	Corrective action is always taken when the leaders are told about unsafe practices
4	In my workplace, managers/supervisors do not show interest in the safety of workers
5	Leaders consider safety to be equally important with production
6	Members of leadership do not attend safety meetings
7	I feel that leaders are willing to compromise on safety to increase production
8	When near-miss accidents are reported, my leaders act quickly to solve the problems
9	My leaders provide sufficient personal protective equipment for the workers

Safety communication with feedback

1	My company doesn't have a hazard reporting system where employees can communicate hazard information before incidents occur
2	Leaders operate an open-door policy on safety issues
3	There is sufficient opportunity to discuss and deal with safety issues in meetings
4	The target and goals for safety performance in my organization are not clear to the workers
5	There are open communications about safety issues in this workplace

Safety policy

1	My leaders explain the safety policy clearly
2	My leaders emphasize worksite safety
3	My leaders have established a safety responsibility system
4	My leaders establish clear safety goals

Safety incentive

1	My leaders reward those who set an example in safety behaviour
2	My leaders praise workers' safety behaviour
3	My leaders have set up a safety incentive system
4	My leaders encourage workers to report potential incidents without punishment
5	My leaders encourage workers to provide safety suggestions
6	My leaders trust workers

Safety training

1	My company gives comprehensive training about health and safety issues to the employees in the workplace
2	New recruits are trained adequately in safety rules and procedures
3	Safety issues are given high priority in training programs
4	I am not adequately trained to respond to emergency situations in my workplace
5	Leaders encourages workers to attend safety training programs
6	Safety training given to me is adequate to enable to me to assess hazards in workplace

Appendix B

Table A2. Items of safety performance questionnaire.

Safety compliance

1	I maintain safety awareness at work
2	I do not neglect safety, even when in a rush
3	I comply with safety rules and standard operational procedures
4	I wear personal protective equipment at work
5	In order to complete more work to get more piece-rate income or measurement of income, I may ignore safety

Table A2. *Cont.*

Safety participation
1   I actively participate in safety meetings
2   I encourage my co-workers to work safely
3   I voluntarily carry out tasks or activities that help to improve workplace safety

Safety accidents
1   In the past three years, I had an accident

References

1. Li, S.; Xiao, L. Statistical Analysis of Domestic Production safety Accidents from November to December 2021. *J. Saf. Environ.* **2022**, *22*, 538–540. [CrossRef]
2. Hystad, S.W.; Bartone, P.; Eid, J. Positive organizational behavior and safety in the offshore oil industry: Exploring the determinants of positive safety climate. *J. Posit. Psychol.* **2013**, *9*, 42–53. [CrossRef] [PubMed]
3. Conchie, S.; Moon, S.; Duncan, M. Supervisors' engagement in safety leadership: Factors that help and hinder. *Saf. Sci.* **2013**, *51*, 109–117. [CrossRef]
4. Dahl, Ø.; Olsen, E. Safety compliance on offshore platforms: A multi-sample survey on the role of perceived leadership involvement and work climate. *Saf. Sci.* **2013**, *54*, 17–26. [CrossRef]
5. Li, G.L. Research on the Safety Leadership of Senior Managers of Coal Mines and the Mechanism of Influencing Dangerous Events. Ph.D. Thesis, Xi'an University of Science and Technology, Xi'an, China, 2017.
6. Li, G.L.; Liu, A.Q. The relationship between safety leadership and leadership style of top managers in coal mines. *China Saf. Sci. J.* **2019**, *29*, 155–160. [CrossRef]
7. Bastos, A.; Sá, J.; Silva, O.; Fernandes, M.C. A study on the reality of Portuguese companies about work health and safety. In *Occupational Safety and Hygiene II*; CRC Press: London, UK, 2014; Volume 1, pp. 687–691. ISBN 978-1-138-00144-2. Available online: https://www.taylorfrancis.com/chapters/edit/10.1201/b16490-122/study-reality-portuguese-companies-work-health-safety-bastos-s%C3%A1-silva-fernandes (accessed on 3 February 2014).
8. Mahmoudi, S.; Fam, I.M.; Afsartala, B.; Alimohammadzadeh, S. Evaluation of relationship between the rate of unsafe behaviors and personality trait Case study: Con-struction project in a car manufacturing company. *J. Health Saf. Work* **2014**, *3*, 51–58.
9. Seo, H.-C.; Lee, Y.; Kim, J.-J.; Jee, N.-Y. Analyzing safety behaviors of temporary construction workers using structural equation modeling. *Saf. Sci.* **2015**, *77*, 160–168. [CrossRef]
10. Amponsah-Tawaih, K.; Adu, M.A. Work Pressure and Safety Behaviors among Health Workers in Ghana: The Moderating Role of Management Commitment to Safety. *Saf. Health Work* **2016**, *7*, 340–346. [CrossRef]
11. Tong, R.; Zhang, Y.; Cui, P.; Zhai, C.; Shi, M.; Xu, S. Characteristic Analysis of Unsafe Behavior by Coal Miners: Multi-Dimensional Description of the Pan-Scene Data. *Int. J. Environ. Res. Public Health* **2018**, *15*, 1608. [CrossRef]
12. Flin, R. Leadership for safety: Industrial experience. *Qual. Health Care* **2004**, *13*, ii45–ii51. [CrossRef]
13. Barling, J.; Zacharatos, A. High performance safety systems: Management practices for achieving optimal safety performance. In Proceedings of the 25th annual Meeting of the Academy of Management, Toronto, ON, Canada, 6–11 August 1999.
14. Larsson, S.; Pousette, A.; Törner, M. Psychological climate and safety in the construction industry-mediated influence on safety behaviour. *Saf. Sci.* **2008**, *46*, 405–412. [CrossRef]
15. Lu, C.-S.; Yang, C.-S. Safety leadership and safety behavior in container terminal operations. *Saf. Sci.* **2010**, *48*, 123–134. [CrossRef]
16. Kath, L.M.; Marks, K.M.; Ranney, J. Safety climate dimensions, leader–member exchange, and organizational support as predictors of upward safety communication in a sample of rail industry workers. *Saf. Sci.* **2010**, *48*, 643–650. [CrossRef]
17. Grill, M.; Nielsen, K. Promoting and impeding safety—A qualitative study into direct and indirect safety leadership practices of constructions site managers. *Saf. Sci.* **2019**, *114*, 148–159. [CrossRef]
18. Kapp, E. The influence of supervisor leadership practices and perceived group safety climate on employee safety performance. *Saf. Sci.* **2012**, *50*, 1119–1124. [CrossRef]
19. Clarke, S. Safety leadership: A meta-analytic review of transformational and transactional leadership styles as antecedents of safety behaviours. *J. Occup. Organ. Psychol.* **2012**, *86*, 22–49. [CrossRef]
20. Fan, Z.Q.; Li, G.J. Research on the effect of active leadership behavior on miner safety behavior. *China Coal* **2016**, *42*, 121–125. [CrossRef]
21. Zhao, S.P. A Study on The Impact of Safety Leadership on Safe Citizen Behavior in Construction Enterprises. Master's Thesis, University of Geosciences, Beijing, China, 2018.
22. Zulkefli, Z. Safety Leadership, Safety Climate & Safety Performance Within TNB's Technical Workforce. In Proceedings of the 9th International Economics and Business Management Conference, London, UK, 26–28 July 2019; European Publisher: London, UK, 2020. [CrossRef]
23. Shafique, I.; Kalyar, M.N.; Rani, T. Examining the impact of ethical leadership on safety and task performance: A safety-critical context. *Leadersh. Organ. Dev. J.* **2020**, *41*, 909–926. [CrossRef]

24. Asad, M.; Kashif, M.; Sheikh, U.A.; Asif, M.U.; George, S.; Khan, G.U.H. Synergetic effect of safety culture and safety climate on safety performance in SMEs: Does transformation leadership have a moderating role? *Int. J. Occup. Saf. Ergon.* **2021**, 1–7. [CrossRef]
25. Cleveland, R. *Safety Program Practices in Record-Holding Plants*; NIOSH: Washington, DC, USA, 1979. [CrossRef]
26. Bryden, R. Getting Serious about Safety: Accountability and Leadership—The Forgotten Elements. In Proceedings of the SPE International Conference on Health, Safety and Environment in Oil and Gas Exploration and Production, Kuala Lumpur, Malaysia, 20 March 2002. [CrossRef]
27. Vinodkumar, M.; Bhasi, M. A study on the impact of management system certification on safety management. *Saf. Sci.* **2010**, *49*, 498–507. [CrossRef]
28. Vredenburgh, A.G. Organizational safety: Which management practices are most effective in reducing employee injury rates? *J. Saf. Res.* **2002**, *33*, 259–276. [CrossRef]
29. Cigularov, K.P.; Chen, P.Y.; Rosecrance, J. The effects of error management climate and safety communication on safety: A multi-level study. *Accid. Anal. Prev.* **2010**, *42*, 1498–1506. [CrossRef] [PubMed]
30. Cohen, A. WITHDRAWN: Reprint of "Factors in Successful Occupational Safety Programs". *J. Saf. Res.* **2013**. [CrossRef]
31. Carrillo, R.A.; Simon, S.I. Leadership skills that shape and keep world-class safety cultures. In Proceedings of the ASSE Professional Development Conference, Baltimore, MD, USA, 13–16 June 1999; ASSE: Baltimore, MD, USA, 1999; pp. 337–344.
32. Hofmann, D.A.; Morgeson, F.P. Safety-related behavior as a social exchange: The role of perceived organizational support and leader–member exchange. *J. Appl. Psychol.* **1999**, *84*, 286–296. [CrossRef]
33. Mearns, K.; Whitaker, S.M.; Flin, R. Safety climate, safety management practice and safety performance in offshore environments. *Saf. Sci.* **2003**, *41*, 641–680. [CrossRef]
34. Probst, T.M. Safety and Insecurity: Exploring the Moderating Effect of Organizational Safety Climate. *J. Occup. Health Psychol.* **2004**, *9*, 3–10. [CrossRef]
35. Zuo, C.X. Research on the Role of Safety Behavior of Managers and Workers in Construction Enterprises on Safety Performance. Master's Thesis, Chongqing University, Chongqing, China, 2015.
36. Oswald, D.; Lingard, H. Development of a frontline H&S leadership maturity model in the construction industry. *Saf. Sci.* **2019**, *118*, 674–686. [CrossRef]
37. Glendon, I.; Litherland, D. Safety climate factors, group differences and safety behaviour in road construction. *Saf. Sci.* **2001**, *39*, 157–188. [CrossRef]
38. Munnich, J.L.W.; Schmit, M.P. Roadway Safety Policy and Leadership: Case Study of Six Midwest States. *Transp. Res. Rec. J. Transp. Res. Board* **2017**, *2635*, 19–27. [CrossRef]
39. Cooper, M.D. *Improving Safety Culture: A Practical Guide*; BSMS: Brighton, UK, 1998.
40. Cao, X.L. A Study on the Impact of Construction Worker Safety Input on Worker Safety Performance. Master's Thesis, Jiangsu University, Zhenjiang, China, 2018.
41. Lu, C.-S.; Tsai, C.-L. The effects of safety climate on vessel accidents in the container shipping context. *Accid. Anal. Prev.* **2008**, *40*, 594–601. [CrossRef] [PubMed]
42. Leung, M.-Y.; Chan, I.Y.S.; Yu, J. Preventing construction worker injury incidents through the management of personal stress and organizational stressors. *Accid. Anal. Prev.* **2012**, *48*, 156–166. [CrossRef] [PubMed]
43. Neal, A.; Griffin, M.; Hart, P. The impact of organizational climate on safety climate and individual behavior. *Saf. Sci.* **2000**, *34*, 99–109. [CrossRef]
44. Al-Refaie, A. Factors affect companies' safety performance in Jordan using structural equation modeling. *Saf. Sci.* **2013**, *57*, 169–178. [CrossRef]
45. Barbaranelli, C.; Petitta, L.; Probst, T.M. Does safety climate predict safety performance in Italy and the USA? Cross-cultural validation of a theoretical model of safety climate. *Accid. Anal. Prev.* **2015**, *77*, 35–44. [CrossRef]
46. Gonçalves, I.; Sá, J.C.; Santos, G. Safety Stream Mapping—A New Tool Applied to the Textile Company as a Case Study. In *Occupational and Environmental Safety and Health*; Springer: Berlin/Heidelberg, Germany, 2019; pp. 71–79. [CrossRef]
47. Cordeiro, P.; Sá, J.C.; Pata, A.; Gonçalves, M.; Santos, G.; Silva, F.J.G. The Impact of Lean Tools on Safety—Case Study. In *Occupational and Environmental Safety and Health II*; Springer: Berlin/Heidelberg, Germany, 2020; pp. 151–159. [CrossRef]
48. O'Dea, A.; Flin, R. Site managers and safety leadership in the offshore oil and gas industry. *Saf. Sci.* **2001**, *37*, 39–57. [CrossRef]
49. Wu, T.-C.; Chen, C.-H.; Li, C.-C. A correlation among safety leadership, safety climate and safety performance. *J. Loss Prev. Process Ind.* **2008**, *21*, 307–318. [CrossRef]
50. Cox, S.; Cheyne, A. Assessing safety culture in offshore environments. *Saf. Sci.* **2000**, *34*, 111–129. [CrossRef]
51. Flin, R.; Mearns, K.; Oconnor, P.; Bryden, R. Measuring safety climate: Identifying the common features. *Saf. Sci.* **2000**, *34*, 177–192. [CrossRef]
52. Zohar, D. Safety climate in industrial organizations: Theoretical and applied implications. *J. Appl. Psychol.* **1980**, *65*, 96–102. [CrossRef]
53. Borman, W.C.; Motowidlo, S.M. Expanding the Criterion Domain to Include Elements of Contextual Performance. *Psychol. Fac. Publ.* **1993**, 71–98. Available online: https://digitalcommons.usf.edu/psy_facpub/1111 (accessed on 13 May 2022).
54. Austin, J.T.; Wolfle, L.M. Annotated bibliography of structural equation modelling: Technical work. *Br. J. Math. Stat. Psychol.* **1991**, *44*, 93–152. [CrossRef]

55. Cheng, L.; Guo, H.; Lin, H. The influence of leadership behavior on miners' work safety behavior. *Saf. Sci.* **2020**, *132*, 104986. [CrossRef]
56. Martínez-Córcoles, M.; Gracia, F.; Tomás, I.; Peiró, J.M. Leadership and employees' perceived safety behaviours in a nuclear power plant: A structural equation model. *Saf. Sci.* **2011**, *49*, 1118–1129. [CrossRef]
57. Griffin, M.A.; Neal, A. Perceptions of safety at work: A framework for linking safety climate to safety performance, knowledge, and motivation. *J. Occup. Health Psychol.* **2000**, *5*, 347–358. [CrossRef] [PubMed]
58. Parker, S.K.; Axtell, C.M.; Turner, N. Designing a safer workplace: Importance of job autonomy, communication quality, and supportive supervisors. *J. Occup. Health Psychol.* **2001**, *6*, 211–228. [CrossRef] [PubMed]
59. Hu, J.; Hu, J.; Hu, H.; Xu, F.; Wang, W. Personalized safety management research for construction workers incorporating personality traits. *J. Saf. Environ.* **2021**, 1–10. [CrossRef]
60. Huang, Y.-H.; Verma, S.; Chang, W.-R.; Courtney, T.; Lombardi, D.A.; Brennan, M.J.; Perry, M.J. Management commitment to safety vs. employee perceived safety training and association with future injury. *Accid. Anal. Prev.* **2012**, *47*, 94–101. [CrossRef]
61. Tam, C.M.; Zeng, S.; Deng, Z. Identifying elements of poor construction safety management in China. *Saf. Sci.* **2004**, *42*, 569–586. [CrossRef]
62. Li, S. The Quantitative Research on Impact of Safety Culture on Safety Performance in Aviation Enterprises. Master's Thesis, Civil Aviation University of China, Beijing, China, 2008.
63. Zhang, Y.M.; Zhang, Y.M.; Qing, Y.F. The scientific way of safety management—Talk about the safety management system of construction en-terprises. *China Saf. Sci. J.* **2000**, 65–68. [CrossRef]
64. Sukiennik, M.; Bąk, P.; Kapusta, M. The Impact of the Management System on Developing Occupational Safety Awareness Among Employees. *Inżynieria Miner.* **2021**, *1*. [CrossRef]
65. Vinodkumar, M.; Bhasi, M. Safety management practices and safety behaviour: Assessing the mediating role of safety knowledge and motivation. *Accid. Anal. Prev.* **2010**, *42*, 2082–2093. [CrossRef] [PubMed]
66. Cavazotte, F.; Mansur, J.; Moreno, V. Authentic leadership and sustainable operations: How leader morality and selflessness can foster frontline safety performance. *J. Clean. Prod.* **2021**, *313*, 127819. [CrossRef]
67. Siu, O.-L.; Phillips, D.R.; Leung, T.-W. Safety climate and safety performance among construction workers in Hong Kong. *Accid. Anal. Prev.* **2003**, *36*, 359–366. [CrossRef]
68. Fan, Y.X.; Fu, G.; Zhu, Y.W.; Wang, X. Overview of the emergence and development of safety management systems. *China Saf. Sci. J.* **2015**, *25*, 3–9. [CrossRef]
69. Zhang, S.; Hua, X.; Huang, G.; Shi, X.; Li, D. What Influences Miners' Safety Risk Perception? *Int. J. Environ. Res. Public Health* **2022**, *19*, 3817. [CrossRef] [PubMed]

MDPI AG
Grosspeteranlage 5
4052 Basel
Switzerland
Tel.: +41 61 683 77 34

International Journal of Environmental Research and Public Health Editorial Office
E-mail: ijerph@mdpi.com
www.mdpi.com/journal/ijerph